The Quantum Brain

The Quantum Brain

The Search for Freedom and the Next Generation of Man

Jeffrey Satinover

John Wiley & Sons, Inc.

Published by John Wiley & Sons, Inc., New York
Published simultaneously in Canada

This publication is designed to provide accurate and authoritative information in re-gard to the subject matter covered. It is sold with the understanding that the publisher is not engaged in rendering professional services. If professional advice or other expert assistance is required, the services of a competent professional person should be sought.

Library of Congress Cataloging-in-Publication Data
Satinover, Jeffrey
    The quantum brain: the search for freedom and the
        next generation of man / Jeffrey Satinover.
      p. cm.
    Includes bibliographical references and index.
    ISBN 0-471-33326-3 (cloth : alk. paper)
    ISBN 0-471-44153-8 (paper : alk. paper)
    1. Cognitive neuroscience. 2. Quantum theory. 3. Brain—Philosophy. I. Title.

    QP360.5 .S28 2001
    612.8′2—dc21   00-033020

Printed in the United States of America

10 9 8 7 6 5 4 3 2 1

To my parents

Contents

Acknowledgments

I am most grateful to the many people who made this book possible and a great pleasure to write. Michael Kellman at the University of Oregon made numerous keen suggestions and comments, which I have incorporated gratefully. But I am especially thankful for his insights into science in general—and about the world we live in. Ramamurti Shankar at Yale took time out of an exceedingly hectic schedule to critique key sections on quantum mechanics. As in his own writing and teaching, his comments were both witty and subtle. Brian Fallon, a psychiatric research scientist at Columbia, read through the entire manuscript in an early stage—no small commitment—and made many extremely helpful suggestions throughout. Many years ago Eric Remole introduced me to neural networks.

All along, it has been important to balance the different requirements of the educated lay and professional reader. Early on, my longtime friend, James Segelstein, helped get me on the right track for the former; Gerd Kunde at Yale likewise highlighted for me some problems I might run into with the latter.

I wish it were possible to have met personally with more of the people referenced herein. I am grateful for the time afforded me early in this project by Stuart Hameroff at the University of Arizona. In an intense but all-too-brief visit, Yael Maguire and Jason Taylor of the Media Laboratory at MIT taught me much—and made me keenly aware of how exciting are the many more things I have yet to learn and didn't know about soon enough to include. Ron Weiss, at work on biological computation at MIT, offered me tantalizing glimpses of how rapidly molecular-level computing technologies are now emerging. Roz Picard, also at the MIT Media Laboratory, and the guiding spirit behind the development of "affective computation," was especially generous with her time, professionally and personally, for which I am most grateful—as is my daughter, Annie, who got an early peek at the future of robotics for her science project. Neither will we forget the sparkling personalities of Cynthia Breazeal and her robotic creation, Kismet. It was a pleasure to chat with them both.

I am grateful as well for the assistance of my agents, Janis Vallely and Martha Kaplan, and for the fine editing skills of Stephen S. Power at Wiley.

I owe a long-overdue thanks to Richard Feynman, who inspired me when I was a child, and who helps keep me young still. If there really is a heaven of the sort he disdained any belief in, he deserves to be there—if for no other reason than because he will so enjoy the last laugh being on him.

My wife, Julie, was a tremendous help, as always. Out of her packed schedule of responsibilities running a three-daughter-plus-one-absent-minded-husband household, she read through the manuscript at all stages, trimming fat and forcing clarification. She is the real inspiration behind the book, and for much else besides—ever since those long breakfasts at the Avalon, in Houston, twenty years ago.

I have dedicated this book to my mother, Sena Satinover, and my father, Joseph Satinover. But a better "thanks" to them is something I've learned: The best gift I can give my own children is what my parents gave to me.

PART ONE

THE QUANTUM BRAIN

INTRODUCTION

THE QUANTUM CRISIS

Our imagination is stretched to the utmost, not, as in fiction, to imagine things which are not really there, but just to comprehend those things which are there.
—Richard P. Feynman, *The Character of Physical Law* (1999)

This book is the story of a revolution that will transform our world and ourselves, one that already has drawn together neurobiologists, psychiatrists, computer scientists, physicists, and mathematicians in an unprecedented competition coordinated more by sheer excitement than by plan. The competition has two interdigitating goals: to achieve an ever more precise understanding of—hence control over—the human brain and to create ever more powerful synthetic brains. We are now approaching both goals with increasing speed, as the race to reach one goal supercharges the race toward the other. We can now see more clearly that both competitions ultimately lead us into the mysterious world of *quantum mechanics*.

When the dust settles, this revolution will, of course, bring about huge advances in science and technology. Scientists at the Massachusetts Institute of Technology and elsewhere, for example, are already planning for a forthcoming Internet that functions as a single, worldwide quantum computer. But the ideas that allow for such a technology represent a revolution in our understanding of ourselves, of life in general, even of God.

ENLIGHTENMENT

Our search to understand the brain has proceeded as has all scientific thought since the Age of Enlightenment: It presumes that there is nothing in the brain (indeed, nothing anywhere in the universe) that is more than *machine*. Neither brain nor mind is anything spiritual or insubstantial; there exists in man no soul, only a collection of physical objects affecting each other via impersonal

3

forces. The unfolding states of these objects, however complicated, are completely determined by whatever states they were in previously: universal billiards without players, payoff, points—or point.

Set in motion once for all time at the Big Bang, particles that later happen to comprise a human brain have no freedom of action whatsoever, neither individually nor as an ensemble. That we think of ourselves as "free," as having "minds" capable of "choosing," indeed, that we even *think we think*: illusion all. What we like to call "will" is at most the inevitable by-product of mechanical interactions of the brain's parts. Illusory "mind" can influence neither what the brain does nor the bodily actions the brain sets into motion.

Poets, mystics, philosophers, and theologians have always insisted that this game of universal billiards has both players and purpose—and their followers have ever fought over who and what these are. But ever since the Enlightenment, science has argued that it is both impossible and unnecessary to know whether the opening break was that of a cosmic Minnesota Fats of incomparable foresight or of a merely comic amateur. In either case, the cue stick lies long abandoned; the player long gone from the shot, the table, the game. The pool hall itself is empty, though the neon lights burn on. In the words of Harvard astronomer Margaret Geller: "Why should [the universe] have a point? What point? It's just a physical system, what point is there?"[1]

For all the millennia that human beings looked at the universe as guided, purposeful, and pregnant with meaning, its operations remained mysterious. But once science arose, the universe swiftly began to yield its secrets. By shifting so wholeheartedly to this point of view that not even life, not even human life, is exempted, science has found itself able to break open even seemingly impenetrable mysteries of mind: learning, intelligence, intuition—all of these can now be extensively understood in wholly mechanical terms. What's more, they are being mimicked by man-made machines. More powerfully than has any prior scientific discovery, the unlocking of the brain seems to confirm the scientists' hypothesis that everything—the mind of man included—is machine.

Steven Weinberg, winner of the 1979 Nobel Prize in physics and an eloquent spokesman for the machine point of view, put it this way: "The more the universe seems comprehensible, the more it seems pointless."[2] And more: "It would be wonderful to find in the laws of nature a plan prepared by a concerned creator in which human beings played some special role. I find sadness in doubting that we will . . . And it does not seem to me to be helpful to identify the laws of nature as Einstein did with some sort of remote and disinterested God. The more we refine our understanding of God to make the concept plausible, the more it [too] seems pointless."[3]

The idea that the entire universe is nothing more than a "physical system"—that is, a machine—unfolding mechanically according to rigid and immutable laws began as the radical heresy of a few brave minds. With this idea as their starting point, they and their followers began to experience an uninterrupted string of successes. There is not a single working medical device or treatment, not a vehicle, not a communications technology, not an industry that isn't based on this assumption. Between the age of Galileo and the end of the twentieth century, the once-radical heresy had become a worldview shared by billions (whether aware or not). So far has this transformation gone that here in America, for instance, where it was once a requirement that a professor in any department in any reputable university be a "man of God," it is now rather an embarrassment should he admit to taking seriously such a thing. The renowned evolutionary zoologist, Richard Dawkins, maintains that anyone who believes in a creator God is simply "scientifically illiterate." By the end of the nineteenth century, scientists believed they had uncovered almost all of the fundamental laws of physics. These laws were all purely mechanical, and out of their mechanical interactions (however complicated-seeming and difficult, in practice, to track these may be) arises every phenomenon we experience at every scale. Living matter itself was understood to be nothing more than an especially complicated factory of molecular machinery.

Needless to say, there were (and are) a great many people who found this vision of reality a horribly bleak one. Most understood little enough about the subtleties of the mechanistic point of view, and of scientific method, to allow them to dismiss it without ever seriously experiencing the power of its claims. But there were a few who did indeed understand its power and hoped that eventually it would somehow be proven wrong. They hoped, in other words, that one day scientists themselves might discover a fundamental law of the universe that was *not* wholly mechanical but in some sense "free." These hopes were rekindled with the emergence of the strange theory of *quantum mechanics.*

QUANTUM LONGINGS

While it is true that at the turn of the last century almost every fundamental physical phenomenon seemed understood, three did resist explanation by the mechanical view of nature. Of course, none of them seemed to have any larger implications about the nature of man, of free will, of life, of God. And to the man in the street they were utterly uninteresting:

- Why does a radioactive nucleus spit out an alpha particle *now* but not *then,* and completely at random? (The mechanical view predicts that there ought to be some kind of regular, internal clock that orders its departure, but there is none.)
- If you heat up any completely black object—like a bowling ball—why is the heat it radiates back at you invisible, and why is it always *heat*, that is, "infrared radiation," and always in the exact same distribution of infrared "colors"? (The mechanical view predicts that the hotter the bowling ball, the more the radiated energy should be as visible light, starting with red, running through the rainbow, and then into the ultraviolet. Note: We're not talking about making the ball so hot that it glows by itself!)
- If you shine light on metal, it will eject electrons—"the photoelectric effect." If the color is right for the metal, *or more toward the blue*, an electron will always be spit out *at once*, no matter how faint the light. But if the color of the light is more toward the red, it doesn't matter if the light

has the intensity of an industrial laser—an electron will *never* be spit out. Why? (The mechanical view predicts that as long as you shine enough light—and there ought to be some minimum—it shouldn't matter what the color is or how intense is the beam.)

These do not seem like earth-shaking quandaries. But when scientists finally did figure out the answers, most of them received a tremendous shock and some had an equally tremendous hope. For the theory that solved these dilemmas—*quantum mechanics*—gave the following Delphic answers:

Nothing at all in the physical universe "causes" an alpha particle to jump out of its nucleus when it does. It just does so, "whenever it wishes." Not only that, it gets out in spite of the fact that the barrier keeping it in is too high for it to get out at all. Were the world really as tidy as scientists had thought, the alpha particle ought no more *ever* be able to get out of its nucleus than could a prisoner in a cell on Alcatraz instantly appear in San Francisco.

Hot, black bodies are black because energy comes in discrete units. This doesn't sound weird, but it's akin to discovering that time only comes in units of hours so you couldn't ever experience or measure time passing during an hour; it would just jump from one hour to next.

A beam of the right color light, no matter how faint, instantly generates electricity in a metal because, *even if the light is spread out like energy, it is at the same time a billiard ball–like particle that knocks an electron out of its orbit.* But if the color is wrong and the energy of the light is too low, it doesn't matter how many particles you throw at the metal, none of the electrons will budge— like throwing thousands of Ping-Pong balls at a bowling ball stuck in a rut.

In the century since it first appeared, quantum theory has created many new dilemmas, solved those dilemmas as well, and all in all proven itself the most successful theory in the history of science. In doing so, it has demonstrated that at its foundation, matter itself does not behave like a machine at all. The very mechanical premises upon which science has been built may be overturned by science itself. This has given some hope that we may find in quantum theory an exit from the dead-end trap of a world that "has precisely the properties we should expect if there is, at bottom, no design, no purpose, no evil and no good, nothing but blind, pitiless indifference," in the words of the eminent evolutionist Richard Dawkins.

Some of these quantum exits take the strange phenomena found at the quantum, subatomic scale and apply them wholesale to human life because of the analogies one can make between, for example, the freedom of choice we believe that we, as people, have and the "freedom of choice" that electrons apparently have. Most serious scientists reject such analogizing because they know enough about how quantum mechanical effects "scale upward" to be convinced that any and all quantum weirdness is long gone by the time we are dealing with aggregates of gazillions of particles large enough to form people.

But as modern biological science has penetrated down into the *subcellular* levels of living matter, and in particular those that constitute the brain, it has indeed begun to encounter the eerie quantum effects that have confounded physicists for a century. These effects are not analogies, they are real, and, as we will see, it is only by considering them that we can begin to understand the building blocks of life. That this is so is not yet widely known to most biologists, but it soon will be. And it will have a dramatic effect on both science and on scientists.

At the subcellular level, matter itself actually looks and behaves (in the words of one physicist) "more like a thought" than like the cogs of a machine. *Nothing in the world that causes the particle to jump*, discovered the first quantum mechanics. But the first premise of science is that *everything happens solely as a result of causes in the world.* "If we are going to stick to this dammed quantum-jumping," complained one of its founders, "then I regret that I ever had anything to do with quantum theory."[4]

Furthermore, if subatomic particles can freely choose to come and go as they please, then perhaps old-fashioned claims as to our own nonmechanical nature aren't so archaic after all: Suddenly, the machinery of brain might prove the illusion, mind and will a more foundational reality. A number of the founders of quantum mechanics wondered out loud whether the ancient mystics might not be right after all: Perhaps there is a Player. Standing apart from the mere "physical system," he everywhere spins the shots, making everything happen *this* way rather than *that.* Wolfgang Pauli thought so: Tongue not wholly in cheek, he simply referred to the so-called exclusion principle, a cornerstone of modern physics and chemistry, as "God."

It turns out, however, that the amount of absolute "freedom" individually available to the bits and pieces of the universe is unbelievably tiny. It amounts to much only on the scale of atomic and subatomic particles. At any scale large enough to be of concern to human beings (e.g., for stuff the size of viruses), the net effect of all that freedom is zero—it just cancels out. Electrons may jump from here to there for no reason whatsoever, but planets don't; nor boulders; nor grains of sand; nor we.

But now the revolution: It appears possible that instead of averaging freedom away as usual, *the human brain,* itself a machine, has nonetheless evolved a unique structure that harnesses subatomic "choice," concentrates it, and amplifies it upward, scale by scale, taking advantage, as we will see, of the strange facts of "chaos." Of all things, it's the machine in our head that lets us transcend our own mechanicality.

Our brains are, if you will, "quantum computers." But they are not of the sort now making headlines. Subtle quantum effects in the brain afford us a capacity we would not otherwise have, yet to make maximum use of such effects our natural brains are now designing even better synthetic ones. These employ quantum principles directly, not, as in the human brain, in subtle and nearly invisible fashion. Some of them will be set free to evolve themselves in Darwinian fashion, hardware and all. But if quantum processes are the source within the human brain of genuine thought—as also of genuine will, intention, and choice—then the quantum computers we are on the verge of designing (or whose evolution we are at least facilitating) may turn themselves into genuine sentient beings. They may have as much intelligence as we have, quite possibly more: There are severe limits to how much quantum weirdness the human brain may employ (and of which sort—a distinction we'll make clear); the lim-

its on how much a synthetic brain might employ are far less severe. Vast, synthetic, self-evolving, superintelligent, and completely sentient computers must
surely sound like pure science fiction, but they are not. Nor are they from some
far distant future.

Furthermore, in the human brain, the amplification of quantum freedom
happens via means that preserve the appearance, at our day-to-day scale, that
we, and our universe, are completely mechanical. All this is terribly confounding to philosophers who seek to understand the world, and human life, in
terms with which they are already familiar. But to really understand where science is leading us may well require a great deal more intellectual discipline and
envelope-pushing than rock-ribbed reductionists, tradition-minded theists, or
New-Age hand-wavers are comfortable with: *"Our imagination is stretched to
the utmost, not, as in fiction, to imagine things which are not really there, but
just to comprehend those things which* are *there,"* in the words of Richard
Feynman, the renowned theoretical physicist.[5] Many scientists speak of a "crisis" brought about by the implications of quantum mechanics and look for
something cleaner, more truly a "mechanics," to replace it and to restore the
austere reductionism of the Enlightenment. Not a few see confirmation of ancient religious notions. But there are some who see something wholly new,
fraught with both fantastic opportunity and terrible risk. If human nature remains true to its history, we may expect both the risk and the opportunity to be
realized, with much gain and much loss. In any event, our understanding of
who we are, where we came from, our standing in the greater scheme of things,
and where we're going, all looks soon to suffer a dramatic change.

1

MIND OUT OF MATTER

THE MACHINERY OF THE FOURTH BRAIN

The brain: An apparatus with which we think that we think.
—Ambrose Bierce, *The Devil's Dictionary* (1906)

The brain is not really a single organ but a vertical stack of them, the newer ones above the older in evolutionary order, with the latest spread out on top. The *neocortex* is a glistening, crumpled sheet of speckled pinkish-gray tissue about a quarter of an inch thick. Stretched out flat like a pelt, it would be about a foot and half wide and two feet long. This pelt itself has six layers, each about twenty-five cells thick. All told, the neocortex is made up of some 10 to 20 billion nerve cells and a stupendously huge number of connections among them, and among the neocortex, the lower brains, and the body. It engenders and guides the vast complexity of intelligent life via processes that are essentially computational—but of a very different sort from those employed by your PC or Mac. The computational machinery of the brain does not execute prewritten programs. *It learns by itself, from experience, and embeds the lessons of experience by reconfiguring its own hardware.*

For many decades, it had been assumed that the cortex must have some kind of immensely complicated "wiring diagram," in which the specific role of each brain cell, and of each connection between brain cells, was laid down according to a fixed plan. But at the same time it was clear that the brain couldn't be wired that way. There were far too many brain cells and they were far too densely interconnected. (Ten billion cells each with 1,000 connections would result in 10 trillion connections.) Furthermore, it was almost impossible to make sense of the overall pattern of these connections: Most seemed jumbled and chaotic; many cells routinely connected back to themselves, creating what surely must be "infinite loops."

Even if there were some kind of plan, the amount of information needed to encode so vastly complex a scheme would exceed the information capacity of the brain itself, let alone the coding capacity of DNA. And anyway, how in

9

the world could it be *physically assembled*—especially at so microscopic a scale—in a mere nine months?

The answer, when it was recently unraveled, surprised everyone. The brain's connections are in fact *largely random.* Yet contrary to what common sense tells us, such a random system *is* capable of intelligence. In fact, it is capable of a higher degree of intelligence—or, more precisely, of a higher *sort* of intelligence—than deliberately designed schemes. It is just this chaotic jumble that gives the human fighter pilot manuevering in combat at supersonic speeds the capacity to do what nearly a quarter billion dollars' worth of high-tech electronics as yet can't. The electronics soon will be able to do likewise, but that's only because its designers are now willing—and able—to let the computer gadgetry train itself, just as human brains do.

Systems such as the brain *organize themselves* into higher and higher levels of information-processing capacity. Global intelligence, we will see, emerges bottom-up from purely local interactions—and by "global," we may even consider the entire *globe.* It is as if the true intelligence guiding the American economy were to reside not in the policymakers of Washington, D.C., but scattered about in the late-night conversations of truckers at truck stops; in the questions of shoppers and shopkeepers' replies; in the friendly chat of neighbors about their rents and mortgages. So, too, the true intelligence of the individual human brain resides not in some central design laid down in DNA but in the happenstance interactions of one neuron with its neighbors (or via long-distance exchanges via the brain's equivalent of the Internet).

It works the other way as well. It is evident that the fundamental building block of a "nervous system" is a single "nerve"—a neuron. But neuroscientists had also long insisted that there exist no smaller information-processing units in complex terrestrial life-forms. Of course, bacteria and other single-celled creatures without nervous systems process information, too, in their way. But these rudimentary processes, it is claimed, have no bearing on human intelligence. We will see, rather, that within each neuron there transpires a complex set of computational events that likewise can be fairly called "intelligent" and that utilizes the same principles of spontaneous self-training as guides the development of the brain as a whole. Up and down the scale, from the smallest elements within the human nerve cell of which study is technically feasible, to the functioning of an entire human society, order emerges from chaos according to the same set of principles.

This sounds almost impossible, but to begin to understand how it works—and to grasp the power we're talking about—we're going to build a miniature, self-learning brain ourselves, and all it's going to take is a few dozen matchboxes and some marbles.

BRAIN IN A MATCHBOX

Tic-tac-toe is a dumb and boring game because there's a pretty simple set of if-then rules that make it impossible to lose—an "algorithm," in computer jargon:

```
1.1.1 If You are X
        Write X in a Corner

   1.2.1 If 0 Responds in the Center
         Write X in Diagonal Corner

   1.2.2 If 0 Responds in an Edge
         Write X in Non-Diagonal Corner

   1.2.3 If 0 Responds in the Diagonal
   Corner
         Write ...
```

The only reason that the game once held any appeal for us is that our parents weren't so cruel as to give us the algorithm ahead of time. But, on the other hand, few of us who now play the game perfectly ever actually sat down one day to figure it out for ourselves. In fact, most of us couldn't just write it down straightaway if asked to for once, even if we never do lose at tic-tac-toe. So how did we *learn* the algorithm? Can we fairly say that we learned it in the sense that we now *know* it? And if we can't really say that we know it, and yet we play according to it anyway, exactly *where* and in *what form* can it be said to exist in our brains? What kind of learning are we talking about?

We learn to play tic-tac-toe like this: We try this and we try that. Some moves are so bad that almost whenever we use them, we lose. So not knowing exactly what we're doing, we try something else—at random. If the results seem better, we try it again. Pretty soon we form (in computerese) a "heuristic," a practical guideline that on average produces better results than having no guidelines or moving at random. This heuristic is something we barely think about directly. We may not even put it into words.

Playing according to our unstated heuristic, we get better and better, modifying the heuristic as we go to avoid losses and achieve victory. Eventually the heuristic "converges" on the complete algorithm, our play becomes perfect, and we give up tic-tac-toe for backgammon, or chess, or commodities trading.

LOSING YOUR MARBLES: GAME AND MATCH TO *HER*

Some thirty years ago, in a *Scientific American* column on machine learning, Martin Gardner described how to build a device out of 25 matchboxes and 116 jellybeans that learns by itself how to play a simple chess-based game he called "Hexapawn."[1] Like tic-tac-toe, Hexapawn is played on a three-by-three board. (See figure 1–1.) It uses only pawns that obey the rules of chess (see box).

The game is quickly analyzed to perfection, which, as Gardner correctly advises, *you should not do*. By *not* analyzing it, you will gain insight into the principles of spontaneous learning. In the thirty years since Gardner's column, these same principles have been extended to puzzles (games, certain complex physical phenomena) that would be impossible to analyze, either because of

FIGURE 1-1   Martin Gardner's Hexapawn. Start of the game.

their sheer complexity (e.g., real chess) or (as in modeling certain physical phenomena, such as fluid turbulence) because they involve insoluble equations. The point is this: Automating a solution doesn't require that first we fully understand the problem (as we can with Hexapawn).

Rules of Martin Gardner's Hexapawn
- Alternating black and white, and in accord with basic chess rules, a pawn may move:
    1. forward one space to an empty square directly in front of it;
    2. diagonally one space to capture an enemy pawn catty-corner to it;
- *En passant* capture (a special chess move), double moves and pawn promotions are either impossible or disallowed.
- White moves first.
- The game is won when:
    1. Any pawn of one side reaches the other's starting row;
    2. One side captures all enemy pieces;
    3. One side blocks the other from further moves.

You can construct the *Hexapawn Educable Robot* (HER) out of 25 matchboxes and 116 jellybeans: 50 red, 46 blue, 16 yellow, and 4 black. Assume that you always make the first move and HER, therefore, only makes even moves. On each matchbox, sketch one of the (only) twenty-five possible board configurations the game can assume following your moves, from beginning to end. (HER will make the second, fourth, and sixth moves.)

As shown in Figure 1–2, represent HER possible moves by arrows numbered from left to right. So long as HER hasn't lost, she will always have as few as one and as many as four possible moves to choose from. You can color the arrows as well to make the coding more obvious. Let the first (leftmost) possible move be represented by a red arrow, the second by blue, the third by yellow, and the fourth by black.

Now, into each matchbox, place *two* jellybeans of the color(s) corresponding to each available move: two red jellybeans if only one move is available; two red and two blue if two moves are available; two red, two blue, and two yellow if three moves are available; two red, two blue, two yellow, and two black if four moves are available.

The game proceeds as follows: Make your first move. Generate a move for HER (from the move options, figure 1–2) by selecting a jellybean at random from the matchbox that corresponds to the state of the board. Make the

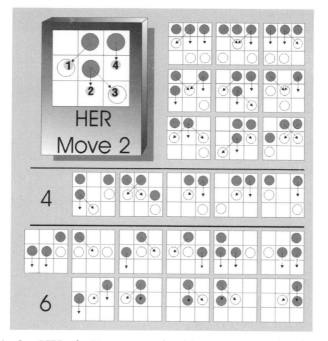

FIGURE 1-2   HER: the Hexapawn Educable Robot. Each matchbox is labeled
with a board configuration and arrows that indicate responses that the robot may
make. Each possible response is represented by colored jellybeans inside the matchboxes. The four colors—red, blue, yellow, and black—are represented in the diagram
by numbers 1 (up) to 4 labeling the arrows from left to right.

indicated move and place the jellybean on top of the box it came from. (Don't
put it back into the matchbox. And don't eat it.) Make your next move. Generate another move for HER, and so on. Since HER moves are entirely random,
Gardner observes that HER ". . . is an idiot." Nonetheless, it is educable—
more precisely, trainable—and it will soon become as smart as you (possibly
smarter), at least in the game of Hexapawn. The key to its progress lies in how
you treat it at the end of a game: the "learning rule." Gardner's learning rule is:

1. If HER won, replace *all* jellybeans in their boxes.
2. If HER lost, eat (or otherwise impound) the bean that came from the
    box that dictated HER *last* move. Replace all others.

Note that under this learning rule, *HER only learns when she loses*; wins
are nice for HER, we suppose, but as far as education goes they are worthless.
(We'll say more about this in a moment.)

HER plays her first game completely at random. But soon moves likely to
be fatal are eaten, and HER's win/lose record improves. That's all there is to it.
You do not need to tell HER what moves are best to make or explain strategy.
By moving at random, HER "learns" from experience. She "converges" on
perfection imperfectly, sometimes getting better, sometimes worse, "two-steps-

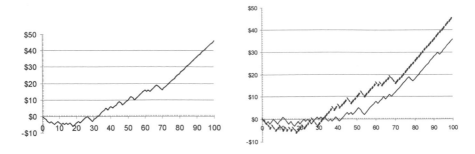

FIGURE 1–3   (a) HER 100-Game Account. Punishment-only training regimen.
(b) Punishment-only (solid line) versus reward-and-punishment (dashed line) learning
rules.

forward-one-step-back" (like us, in most things), but in the end, if you were playing HER for a dollar a game, she would wipe you out, as figure 1–3a illustrates.

There is no need to teach HER strategy. There is consistent feedback from the environment: Losses have consequences different from wins. There is nothing creative, insightful, self-reflective, or canny about the "intelligence" acquired by the device; no "critical thinking" is needed. Indeed, the learning process is rote and wholly mechanical. Major breakthroughs (lengthier uptrends) often occur just after a string of losses. Finally, HER not only tolerates randomness, she takes advantage of it. *Were she somehow to avoid blind forays that prove wrong from time to time, she wouldn't learn a thing.* Nor would we.

In my years as a psychiatrist, I have treated many highly "successful" individuals who have floundered upon confronting setbacks, considering themselves "failures." I have learned the following lesson: *Truly* successful people, the most successful, *have a very checkered track record,* peppered with what they consider many serious "losses." Success for them is defined not by any external, objective standard unvarying from one person to the next but rather by that level of accomplishment that the individuals themselves experience as unequivocally satisfying (and that often happens to, but need not, coincide with what others consider successful, too).

Invariably, those with near-perfect records, outstanding by "objective" standards, suffer from two afflictions: They are excruciatingly sensitive to failure (which is why when confronted with it, should they happen to be, they fall into the tailspin that brings them to my office); they consider themselves never really to have fulfilled their own potential, even when not in crisis. *And they're correct,* contrary to all the kind-hearted, humane and utterly useless self-esteem–building that constitutes a substantial part of "therapy" nowadays.

In reality, a decent exposure to failure, over a long enough time, not only inoculates us against emotional collapse, it allows us to try things that expand our reach: If you can't fail, you can't succeed. The steepest learning curves (measured against a standard that varies from person to person, however) happen when serious loss is allowed and not avoided. When we avoid loss, we still may do well by the world's standards, but *we know we are being cowardly.* We therefore can't experience the joy of genuine accomplishment—and we shouldn't.

Of course, we do not learn *only* from failing; success surely helps, too. So, having learned something about human behavior from HER, what about applying what we know about people to training our machine? Let's compare the original, losses-only regimen to a new one where we also reward HER after *wins* by *adding* jellybeans to the last box. (See Figure 1–3b.)

We see that if the goal is to train HER to achieve perfect play as soon as possible, then a pure punishment-based learning rule works better than reward and punishment. But if the goal is to maximize the number of games won in a relatively short tournament (between twenty and thirty games), then a reward-and-punishment regimen is better. This relationship will not hold in every circumstance—it depends on the specifics of the learning machine and its "learning rule." But it does highlight an important principle: The "best" method of education can only be defined with respect to an exact definition of what the goal of education is. Many a modern educational "improvement" is not that at all, but rather a tacit redefinition of educational goals. "Reward-only" methods, for example, which are popular today, do not necessarily produce "better" students by most standards, but they do seem to produce students who consider themselves better. (A reward–only regimen will teach HER to play Hexapawn well only very slowly, and HER will never become a perfect player. But she never need sacrifice any jellybeans.)

HOW HUMAN?

How convincing are the analogies between HER and human learning? A Montreal high school mathematics teacher told two students that, while sitting in separate rooms, they were going to play Hexapawn against each other, relaying moves through an intermediary. He asked them to comment on the play. In fact, each was playing against a HER.

> "He took me, but I took him, too. If he does what I expect, he'll take my pawn, but in the next move I'll block him."
> "Am I stupid."
> "Good move! I think I'm beat."
> "I don't think he's really thinking. By now he shouldn't make any more careless mistakes."
> "Good game. She's getting wise to my action now."
> "Now that he's thinking there's more competition."
> "Very surprising move. Couldn't he see I'd win if he moved forward?"
> "My opponent played well. I guess I just got the knack of it first."
>
> The teachers and his other students watched through a mirrored glass:
> With much confusion and muffled hilarity we in the middle tried to operate the computers, keep the games in phase, and keep the score. . . .

When the students were later brought face to face with the machines they had been playing, they could hardly believe that they had not been competing against a real person.[2]

May we therefore claim that from an inanimate, random-acting, and so fundamentally stupid collection of household objects something that looks suspiciously like intelligence and human learning has spontaneously emerged?

Well, maybe not. There is obviously thought and intelligence *behind the way the matchboxes are "designed" to learn, in how cleverly they take advantage of randomness;* in the specifics of the learning rule, for example. But suppose we set up an extra matchbox whose jellybeans would randomly code not for one of a set of possible moves but for one learning rule from a set of many different rules. In fact, we might have a whole set of such extra matchboxes (call it the "superset"), each of which selected one *word* in a training rule. Before starting a Hexapawn training session, we'd first use the superset to create a training rule for that session.

Now, since the words of the rules start out randomly ordered, at the beginning, HER is almost certainly going to learn nothing—its rules will be senseless. But if we limit each training session to, say, 100 games, then forcibly start over (HER dies, as it were, and is replaced by an infant version), and if we have a decent "super-learning-rule" that determines what we do with the jellybeans in the superset of matchboxes, then after a certain number of training sessions—say, 100 sessions of 100 games each—we will not only have arrived at a learning rule, we may arrive at close to the *best* learning rule. HER would not merely learn how to win at Hexapawn, she would learn how to best learn how to learn to win at Hexapawn. ("Best," of course, will be determined by what *we* choose to reward and punish in the superset.)

What have we accomplished? Well, *we* are no longer implementing a learning rule of our choice, nor are *we* rewarding and punishing HER: The superset is doing all of that. But so what? There is obviously thought and intelligence *behind the way the superset is "designed" to learn, in how cleverly it takes advantage of randomness;* in the specifics of the learning rule applied to *it,* for example. We've just pushed our involvement back one level, so it's a bit more hidden. But why stop there?

HERCULES, a.k.a. ROBOTIC HUSBANDRY

We now create multiple "copies" of HER supersets and have the various learning rules they come up with *compete against one another.* Losing supersets would be removed from the competition; the rules of those that remain we would cut up and shuffle, at random, to create new "generations" of HER supersets—for example, the punishment half of one rule might be paired with the reward half of another.

We would then set *these* to competing both against one another and against their "parents." (Alas, life in the HER world is nasty, brutish, and short.) In this way we would not only stumble upon good ideas via random variation, we would *evolve* better and better rules. (The mathematical formalization of this kind of solution-seeking is called a "genetic algorithm"; the mathematical formalization of computer programming optimized this way is called "evolutionary programming.")

Of course, we'll also want to introduce a certain rate of "random mutation" into the pairing of learning rules to allow for the emergence of wholly new features, not just combinations of existing ones. We'll also breed rules of differing "fragilities"—that is, of differing average frequencies of random mutation of learning rules, so as to evolve a mutation rate that optimizes the evolutionary process itself.

In short, HER will turn out merely to be a "cell" within a larger machine that we can call HERCULES: Hexapawn Educable Robotic Colony Undergoing Learning and Evolution via Selection.*

The tissue of the human cortex, the topmost brain, is structured in just this way—almost. It is itself a gigantic set of learning machines, with supersets and subsets as we've described (cells and collections of cells and collections of collections of cells). But there is one big difference: the *parts of which each learning machine is composed* are themselves miniature learning machines; and the parts of which these are composed are, too. It's as though the matchboxes and jellybeans in a HER were each, themselves, learning machines composed of parts which are learning machines, composed of parts. . . . Were this so, a big part of the influence exerted on a given HER would proceed not from the *outside* (from us, at a larger scale), but from the *inside* (at a smaller scale). This puts rather a twist on the idea that if we keep pushing back, eventually we'll find the designer.

To discover what it is we do find, let's turn to a more precise description of how "matchbook games" play themselves out in the human fourth brain, spontaneously giving rise to intelligence, language, creativity, genius—and madness, too.

---

* Certain readers may be offended by the great prominence given a female-seeming example—HER—in the preceding sections; other readers when it turns out that the next larger-scale entity of which HER is a part seems male—HERCULES. Such readers may wish to console themselves with a belief that the *ultimate* Hexapawn superset—should it exist—has a gender of the wished-for variety and cares how you refer to it.

2

OPENING THE MIND'S EYE

*"May not one be admitted to inspect the machinery of Wisdom?
I feel curious to know how thoughts—real thoughts—are born."*
—Lady Blandish in *The Ordeal of Feverel*, by George Meredith

In the seventeenth century, shortly after the invention of the telescope, the great mathematician, computational scientist, and religious philosopher Blaise Pascal undertook a careful study of the motion of the planets. One day he mentioned to his mother that, like the moon, Venus goes through phases. Pascal handed the device to his mother to take a look at the planet, which at that moment happened to be in a crescent phase. Though not herself a scientist, Pascal's mother was an intelligent woman with a keen interest in science. She had never before looked through a telescope and when she did so, she promptly observed that "Yes, of course Venus has phases, but it's the reverse of what it should be."

Pascal was startled. "How do you know that?" Telescopes did indeed invert images, but how would his mother know what the *proper* phase should look like, having never had her hands on a telescope before—indeed, how could she know that Venus had a phase at all? The phases of Venus were only discovered once telescopes were available.

"Because I can also see it under my naked eye."

Pascal had no particular response to this assertion. He certainly could not see Venus as anything more than a large point of light.

Was her vision better? Was she more observant? Three hundred years later we would find the answer. It was not her vision, it was the human retina. The retina is not really a "part" of the eye at all. It is rather a physical extrusion of the fourth brain itself *into* the eye. And like other brain matter, the retina therefore is organized not merely to gather light but *to intelligently process visual data*. It is a vastly powerful, sophisticated, and complex *pattern discriminator and classifier*. Insofar as one may refer to "machines that think," the retina is an enormously subtle thinking machine—a biological computer of great power and beauty, packed into an impressively small size. Imagine, if you will, a desktop computer about the size and thickness and delicacy of a rosepetal. And what limitations it has are now understood to lie far *below* the threshold of acuity needed to detect the phases of Venus with the unaided human eye—perhaps with the unaided eye of any terrestrial creature.

For the answer to Pascal's query we can largely credit the efforts of one flamboyant scientist, Frank Rosenblatt. Rosenblatt was a pioneer of *neural computation*—the construction of electronic devices that process information not according to "top-down" rules of logic but by mimicry of the "bottom-up" wanderings of nature. His prime device—the Perceptron—today sits in the Smithsonian, next to the von Neumann computer that made possible the atom bomb. Of the significance of the latter, very many are at least dimly aware: Every PC in the world was fathered by it; but of the former, very few. Indeed, there remains considerable irritation in some quarters that the Perceptron is there at all. But it is the father of the computer of the future, and it was modeled on the human retina.

ROSE PETALS

Rosenblatt had developed a fierce and precocious fascination with the quest for machine-based intelligence, just then beginning, in the late 1940s. It seemed obvious, in those days, that "artificial intelligence" could best be developed by studying and copying natural intelligence. But the task proved more difficult than anticipated. Within twenty years, the burgeoning science of computation would abandon the biological template to develop the kind of machines that today sit on almost every desktop.

Like many young scientists to follow, Rosenblatt's passionate quest for biologically informed computation was fueled by a fantastically clever and influential article written by two pioneers at the Massachusetts Institute of Technology. In 1943 Warren McCulloch and Walter Pitts published a piece in the *Bulletin of Mathematical Biophysics* entitled "A Logical Calculus of Ideas Immanent in Nervous Tissue." With mathematical certainty, the article showed that a collection of nerve cells was not only capable of computing, but given how individual neurons behaved, and how they were connected to each other—with a lot of randomness—they would *necessarily* compute.

In particular, they showed that the "ideas" embodied in a collection of neurons were not explicit, as in high-level human languages, but implicit ("immanent"), carried by the collection of neurons as a whole in much the same way that in matchbox Hexapawn, no individual matchbox embodies the "idea" of the game, but, once trained, the entire assemblage of matchboxes does.[1] McCulloch and Pitts's paper is little known to the world at large; to the computational science community it is universally known and admired, and has been credited with spurring the entire computer revolution. But the first use to which it was put was in creating a *neural network*: a densely interconnected set of elementary processing elements that, as a whole, could spontaneously develop powerful intelligence.

Fifteen years after McCulloch and Pitts's seminal paper, Frank Rosenblatt created the Perceptron, a neural network based on the retina. By repeatedly processing information in network fashion—also called *distributed* or *massively parallel processing*—a group of even relatively simple neurons can acquire a fantastic capacity for discrimination (as the HER matchboxes acquired strategic ability). This is why the naked eye can in fact detect the crescent shape of Venus.

The elements in a neural network, and the neurons in the retina, operate roughly like this. An incoming signal (say, the local intensity of light) is stimulated by a detector neuron. It transforms the intensity into an electrical signal of a corresponding strength, which it then distributes to many other neurons. There are many such detector neurons, and they all distribute their individualized output to the many other neurons. Each adds up its inputs and similarly converts the net result into a corresponding output. In short, each of many neurons receives many different inputs from which each synthesizes a single output to distribute to many others—hence, "massively parallel."

If that's all there were to it, nothing would happen: Such a system couldn't learn. But the network of neurons has an additional mechanism that is the equivalent of HER jellybeans. The connections between neurons are themselves of varying strength (usually called "weight"): Depending on their weights, the connections either enhance the signal they are transmitting or diminish it. Since at the beginning these weights differ at random, neural nets initially scramble any incoming signal and put out noise. But in a living nervous system, the system itself modifies the weights in light of experience: Connections that frequently carry signals, especially strong ones, are themselves strengthened; connections that infrequently carry signals, or mostly weak ones, are themselves weakened, a mechanism first outlined in 1949 by neuropsychologist Donald O. Hebb in what has become a landmark book, *The Organization of Behavior: A Neuropsychological Theory.*[2]

It's almost like a statistical reasoning process: "Hmm. It seems to happen again and again that whenever interest rates are lowered, stock prices go up. From now on, whenever interest rates go down, I'm going to get excited about the stock market, even though I have no theory whatsoever as to why the two events should be connected. They just are—in my mind, at least."

Over time, the network diminishes connections that contribute mostly "noise" and bolsters connections that for the most part "work." Eventually the network "memorizes" the incoming pattern as a specific distribution of varying connection strengths, in the same way that HER "memorized" the strategy of Hexapawn as a distribution of varying matchbox contents. Furthermore, as long as the density of interconnections among neurons is sufficient, the connections themselves can be random: No "wiring diagram" is needed, just as HER needs no logical instructions in strategy.

In an artificial device, we modify the weights by hand, from the outside. *Biologically plausible* schemes such as "Hebbian learning" were the first step toward an understanding of how *local, lower-level systems* can influence the global behavior of a composite whole, without external supervision or human tinkerers.

This relatively simple process conforms to the actual structure of networked biological neurons. These typically have multiple short input channels called "dendrites" and a single long output channel called an "axon." The axon then branches out to reach at least one dendrite on many other neurons.

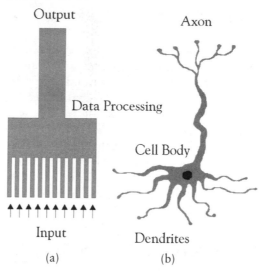

FIGURE 2-1   (a) Schematic of an artificial neuronal processing element showing correlation with the structure of a typical biological neuron. (b) A biological neuron showing four basic structural elements that determine data flow and processing.

The structure itself suggests that the body of the neuron combines multiple inputs to produce some sort of overall total (or average), which it then puts out. Figure 2–1 shows how artificial processing elements are modeled directly on this biological structure.

YES, WE HAVE NO BANANAS

The simplest possible Perceptron consists of but a single processing neuron that performs a yes/no classification of data it receives from a set of detector neurons. For example, if a fruit is "banana," it puts out (+1); if "no banana," (–1). To do this, it must receive and process ("detect") features of the object. There might be two detector neurons, one that emits a signal only when it "sees" the color yellow and a second that emits a signal when it "sees" long objects. As various fruits are presented to it, it slowly gets better and better at putting out a (+1) only when presented with a (not overly ripe) banana.

Recall that when we began training HER, the robot selected moves at random. The Perceptron likewise at first "guesses" at random whether an object is or isn't a banana, with about a fifty-fifty chance of being correct, *because initially its weights are set at random.* To train the Perceptron, we present it with many different bananas and nonbananas. Learning occurs as we adjust the weights in response to whether the Perceptron wins or loses the game of "Am I a banana?"

But here's the challenge: Just how, precisely, should we adjust the weights so as to guarantee that after a time the Perceptron will correctly learn how to discriminate between bananas and not bananas? That is, "What is the correct learning rule?"

In a series of papers and books published between 1958 and 1962, Rosenblatt came up with an elegant scheme for adjusting synaptic weights so as to ensure learning and, even more important, provided a mathematical proof—the *Perceptron convergence theorem*—that Perceptrons using his adaptation scheme will always *converge*, that is, the weights will always achieve a set of values such that the Perceptron will classify correctly.[3] The scheme basically amounts to: (1) do nothing if the Perceptron guesses correctly; (2) if the Perceptron says something is a banana and it isn't, decrease the weights by a small percentage of the input values they multiply; (3) if the Perceptron says something isn't a banana and it is, increase the weights by a small percentage of the input values they multiply.

OUT OF MANY ONE (E PLURIBUS UNUM)

An obvious next question is "What if there are more than just two classes to the data?" What if the Perceptron is required to discriminate among bananas, *apples*, and neither?

In fact, the extension of the single Perceptron to a Perceptron layer—once again modeled on the retina—is both subtle and straightforward. You need only line up as many Perceptrons as you need side by side and have them all receive data and feed it forward "in parallel" (hence, "parallel processing"). For each additional class, you simply add one more Perceptron, as in figure 2–2.

With the convergence theorem demonstrated and the extension of the single McCulloch-Pitts neuron to a layer of such neurons modeled on the retina, a tremendous amount of enthusiasm was generated for the Perceptron for the following reason: A large and rapidly growing class of problems in engineering, optics, biology, physics, and the social sciences were being expressed by a large number of simple, straight-line formulas as a substitute for a small number of very complicated equations. It turned out that large numbers of

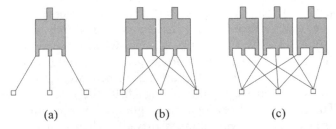

(a)   (b)   (c)

FIGURE 2–2 Perceptrons in parallel. Three small square neurons form the bottom, input layer. Each neuron is sensitive to one feature of whatever is being presented. The next layer—the "processing" layer—is composed of one, two, or three neurons in examples (a), (b) and (c), respectively. The output emerges from the top of this layer. (a) Single Perceptron with three feature inputs, one binary (on/off) output. (b) Two Perceptrons linked in parallel. Three feature inputs to both, two binary outputs. (c) Three Perceptrons linked in parallel. Three feature inputs to all three, three outputs.

straight-line equations could be converted into, and therefore solved by, the behavior of Perceptrons. It began to seem possible that a complex enough Perceptron layer could converge on the solution to almost *any* problem. The machine would *learn* how to solve these, even if human beings couldn't.

But another—and in critics' view, a more important—factor in the enthusiasm for the Perceptron was the personality of Rosenblatt himself. "He was a press agent's dream," in the words of one scientist (which translates precisely as "every scientist's nightmare"), "given to steady and extravagant statements about the performance of his machine": "A real medicine man. To hear him tell it, the Perceptron was capable of fantastic things. And maybe it was, but you couldn't prove it by the work Frank did."[4]

Soon the number of researchers turned off by the style of the Rosenblatt "road show" grew larger than the number turned on by a full appreciation of what he had discovered. And many high-powered artificial intelligent experts with a sophisticated mathematical background began to suspect that the Perceptron was mostly smoke and mirrors—that it could do a few things well but certain important things not at all. Chief among Rosenblatt's critics was Marvin Minsky, who would soon claim that that he could mathematically prove the Perceptron incapable of anything "interesting." Doubt about the Perceptron turned into anger as Rosenblatt's reputation grew: "Leon Harman, who worked on the von Neumann machine at the Institute for Advanced Studies at Princeton, and who describes himself as perhaps the first computer operator, still seethes about walking into the Smithsonian and discovering that beside the von Neumann machine, which well-deserved to be there, stood a Perceptron, sharing floor space as if it were equally important. . . ."[5]

Soon the premise that we can learn anything about computation from biology, or about how living systems "think," acquired a taint, with the same odor of hype and dilettantism that today clings to pop quantum mysticism: "[Minsky] doubts that we'll ever learn much about brain operation from studying electronic hardware, and believes that the really interesting and potent things the computer in our head does are inscrutable."[6]

Somehow, the critics had forgotten von Neumann's own opinion:

> It has often been claimed that the activities and functions of the human nervous system are so complicated that no ordinary mechanism could possibly perform them. . . . It has been attempted to show that such . . . functions, logically, completely described, are per se unable of mechanical neural realization.
>
> The McCulloch-Pitts result puts an end to this. It proves that anything that can be . . . completely and unambiguously put into words is ipso facto realizable by a suitable finite *neural network*.[7,8]

Von Neumann's prophetic words went unheeded. Before the 1960s fully got under way, critics of the Perceptron evolved into critics of *any* kind of biologically based machine computation. Their formal criticisms, as we will soon see, appeared devastating, and the direction of computer development was drastically altered, becoming what we know it as today.

After his brief romance with fame, if not the fortune that today attends computational triumphs, Rosenblatt, along with his Perceptron, was forgotten. Nonetheless, some did have second thoughts about his quick dismissal. A critic *and* admirer of Rosenblatt put it this way: "He did irritate a lot of people, but he also charmed at least as many, and I count myself among them. Just when you were thinking that Frank didn't have another trick up his sleeve, along he'd come, and he'd be so darn convincing, you know he just had to be right.[9]

And he was. But by the time he was vindicated, a quarter century had passed, and Rosenblatt, still in his prime, was dead.

3

DEATH AND BIRTH

In the days when [artificial intelligence expert Gerald] Sussman was a novice, Minsky once came to him as he sat hacking at the PDP-6.

"What are you doing?" asked Minsky.

"I am training a randomly wired neural net to play tic-tac-toe," Sussman replied.

"Why is the net wired randomly?" asked Minsky.

"I do not want it to have any preconceptions of how to play," Sussman said.

Minsky then shut his eyes.

"Why do you close your eyes?" Sussman asked his teacher.

"So that the room will be empty."

At that moment, Sussman was enlightened.

—jargon, node: Some AI Koans
http://igwe.vub.be/links/jargon/s/SomeAIKoans.html

"Many in computing remember as great spectator sport the quarrels Minsky and Rosenblatt had on the platforms of scientific conferences during the late 1950's and early 1960's."[1] But the Minsky-Rosenblatt debate was more than mere sport. It was one of the great fights of twentieth-century science, informed by brilliant engineering savvy, even more brilliant mathematics, and all the imperfections of human personality.

The sides of the debate were clear: According to Rosenblatt, biologically inspired computation could do almost everything; according to Minsky, it could do almost nothing. The debate heralded a new era of artificial brains—and the rise of a new, competitive aristocracy of natural ones. When, in the late 1960s, Minsky seemed at last to have won flatly, neural-based devices withered and computation developed into what we know it as today: so-called expert systems—utterly dumb mechanical processes that merely implement, but with blazing speed, the "top-down" instructions of a human expert. It was a model that comported nicely with the worldview of that era, in which social engineering was the rage and the worship of experts of every stripe was rapidly accelerating.

MINSKY

After leaving the Bronx High School of Science, Rosenblatt's own alma mater, Marvin Minsky went to Andover and then Harvard. There he promptly obtained his own laboratories, as an undergraduate, in no less than three separate departments: physics, biology, and psychology—the three areas he thought he would need to understand the brain or, since "the brain happens to be a meat machine," as he put it, to build his own.[2] Like Rosenblatt, he, too, as a youngster, was at first entranced by the biology-based computational potential of the McCulloch-Pitts model.

Soon Minsky had wired together some three hundred vacuum tubes in an amazingly sophisticated neural network that simulated the behavior of four rats progressively learning to avoid each other as they navigated a maze—a considerably more complex learning task than a Hexapawn robot. His net incorporated randomness and worked well, but Minsky was dissatisfied with the theory he developed to explain *how* it worked.

Minsky abandoned the contraption to Harvard and proceeded to Princeton, where he invented the method of so-called reinforcement learning, a subject that continues to receive attention to this day. Having devised a somewhat more satisfactory theory of his net, he obtained the Ph.D. in 1954 with a superb neural network–based dissertation. He was soon a rising star at MIT.

But Minsky was beginning to suspect that adaptive, biology-based computation would prove a dead end. Or maybe he was just bored: "I always sort of look forward to [abandoning a project] expectantly. It's very nice, see, because you don't have to finish that thing."[3] In any event, he soon became the arch-foe of neural networks, the Perceptron included. And Minsky's opposition to biological computation was all the more effective precisely because he had such impressive early neural network research under his belt. He began to design and champion logic-based expert systems instead. Like Minsky's own personal "meat machine," these were exquisitely precise—in contrast to Rosenblatt's flamboyant one. Under his influence, the term "artificial intelligence" came explicitly to mean *not* Perceptron-like neural nets.

TAKING IT ON THE BRAIN

Neural nets are based on the premise that naturally occurring neurons and naturally occurring nets of such neurons provide the best possible model for information processing, precisely because over untold millennia of experimentation nature had evolved it. But even if that's how our *brains* generate intelligence, is such "distributed parallel processing" really the way we learn? Certainly not always:

Socrates is a man. All men are mortal. Therefore . . . ?

. . . Socrates is mortal. This we conclude by logic, not by trial-and-error "convergence." Our teachers do not simply whack us repeatedly over the knuckles with a ruler until we haplessly mouth the correct sequence of terms, in a desperate random search for surcease of pain.

Furthermore, neither Minsky nor anyone else had ever seen a pile of junk evolve itself into a thinking machine; most of it had to *be created by clever human beings in the first place*, whether it was constructed out of matchboxes or vacuum tubes, just as he had done. The most you could do with any *actual* learning machine to make it appear spontaneously intelligent is to push one level or more into the background the fact of a human designer—as we did with our nested supersets of Hexapawn. Eventually Minsky concluded that it was a foolish conceit to treat "randomly wired neural nets" as genuinely random at all: They were *designed* that way by people. Not seeing this fact was as deliberately self-deluding as closing one's eyes to make the room go away.

True, there did seem to be *one* example of a pile of junk that had evolved itself into a thinking machine: the human "meat machine" itself. But that had taken *billions of years* and the resources of an entire planet. So Minsky abandoned neural nets and began to design machines that think by doing what we do when we think well: We head to the answer in a precise sequence of logical steps, not in a drunken stagger.

In 1961 Minsky met a like-minded theoretician named Seymour Papert at a conference in England. They decided that together it would take them a year or two to settle the Perceptron controversy once and for all. Minsky brought Papert back to MIT, and they began to analyze with mathematical rigor exactly what it was a Perceptron could and couldn't do.

But as they tied up one loose end here, another appeared over there. Again and again they delayed publication. Finally, in 1969, they were ready. Prepublication copies of *Perceptrons: An Introduction to Computational Geometry* made an instant sensation in the world of computational intelligence. The tone of what was to come was set on page 4, where they offered their general assessment of the huge literature on Perceptrons: "Most of this writing," they asserted, ". . . is without scientific value."[4] "My God, what a hatchet job," marveled one professor of electrical engineering at Stanford.[5]

Minsky was promptly invited to deliver a preliminary version of *Perceptrons* to the prestigious American Mathematical Society. He had already established himself as a formidable speaker, heavy horn-rimmed glasses and plastic pocket protector notwithstanding:

Minsky's diction is as precise as a trained actor's, his knowledge nearly universal. He shares with the rest of the founding fathers of AI an omnivorous appetite for experience and knowledge—of music, medicine, science fiction, history, engineering, mathematics, politics, futurism, fantasy. . . . Watching Minsky slice up a colleague whose ideas seem ill-considered to him is as much terror as sport, and though the voice is well-modulated, the hands are fidgeting and tense, a sign of his emotional involvement.[6]

By 1969 the challenges of making neural nets work properly were found daunting by many others; skepticism about Rosenblatt's larger claims was beginning to spread; Department of Defense (DARPA) funding had always been subject to unexpected shifts in fashion, and one was beginning with respect to computing (because of how much Minsky had accomplished already with expert systems). But the Minsky-Papert book was a tour de force demonstration, on the most rigorous of mathematical grounds, of all the truly crucial things Perceptrons *would never be able to do*. In particular, certain elementary logical operations, essential for *any* computational method, were beyond them. For example, a Perceptron could not even handle so simple a pattern discrimination task as mapping data into regions, if the boundary between the regions was anything other than a perfectly straight line. Closed regions within regions (like a county within a state) were yet further beyond its capacities—in spite of the Perceptron's being touted as a nearly universal "pattern-recognizer." Listeners understood that they were hearing the coup de grâce.

This limitation of the Perceptron is extremely profound for the following reason: The basic elements of *all* computing can be shown to be built up out of a small set of elementary logical operations such as: NOT (as in "Dogs are NOT men") or AND (as in "Socrates is a man AND mortal").

The other important logical operation is OR, but here we need to be careful, for there are actually two of them, OR and XOR, and they're different. OR ("or") means "A or B or both," as in "If Socrates is a philosopher OR a man, he's mortal." XOR ("exclusive or") means "either A or B but not both," as in "Socrates is either mortal or immortal." Both may not be true, nor may both be false.

Let's express this in more general terms. Suppose we find only "black" or "white" as possible hair colors and as possible skin colors in a certain country (Penguin-land, perhaps). Then HAIR XOR SKIN means that we assign to one group those people who have *either* white skin or white hair, but not both; and we assign to another group those who have *both* white skin and white hair, or both black skin and black hair. The first group we will then label "White," the second "Black." (Two values are being condensed into one according to the XOR rule.) Figure 3–1 is a table of the possibilities.

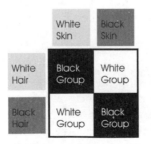

FIGURE 3–1   XOR: Logical "exclusive or" creates groups that are not contiguous.

There is no way of drawing a single straight line to separate the "White Group" from the "Black Group." XOR, one of the most basic elements of logic, is therefore beyond the Perceptron. Once this most basic weakness was exposed, the judgment was sure and swift: Perceptrons are utterly useless, the farthest thing imaginable from a "brain."

Would it perhaps help to add additional layers to the Perceptron, somewhat like creating an approximate curve out of many straight-line segments? Minsky and Papert offered their intuitive judgment that ". . . the extension to multi-layer systems is sterile"[7] (by which Minsky means "will generate nothing of value").

Minsky's youthful fascination with biologically based computation was finished, as was any further need for competition with Rosenblatt. Over the next few years Minsky would become the most eminent representative of genuinely *symbolic* computation. Funding agencies' fascination with biologically based computation was finished as well, especially the crucial support that came from DARPA. The wave of dollars sloshed away from the ever more isolated islands of neural network research and toward a new world of computation: the broad shores of Minsky's artificial intelligence.

A few years later, Rosenblatt, out sailing alone on a lake, drowned. Rumors persist to this day about his state of mind at the time, and the reasons for and means of his death, but his closest colleagues insist it was a freak accident. Hit by a boom he didn't see coming, he was knocked overboard. His epitaph? *Without scientific value.*

DEATH XOR RESURRECTION:
THE SECOND COMING OF NEURAL NETS

But not everyone was convinced that the Perceptron, and other bottom-up systems, had suffered the coup de grâce the majority was convinced it had. In a few university laboratories around the world work continued. It was hampered by lack of funding but also stimulated by it—those who pressed on were thoroughly convinced of its merits, independent-minded and highly motivated.

Bernard Widrow at Stanford, for example, had developed a Perceptron-like device he called the "Adaline," short for "adaptive linear element," the "linear" referring to the straight-line limitation we discussed above. When he received a copy of *Perceptrons,* he "figured that [Minsky and Papert] must have gotten inspired to write that book really early on to squelch the field, to do what they could to stick pins in the balloon." All the authors had done was to define neural computation so narrowly "that they were able to prove that it could do practically nothing." In particular, Minsky's rather lofty assertion—he had not tried to prove it—that multilayer neural nets were "sterile" was flatly wrong. The problem of "linear separability," as it was called, of the XOR, had long ago been solved in just this way, Widrow points out, but no one was paying attention.[8] He himself had also developed "Madaline," built of *many* Adalines.

OUTSIDE THE BOX

"All this talk of space travel is utter bilge." So insisted the British Astronomer Royal, Richard Wooley, in 1956, less than a year before the Soviets launched Sputnik. Thirteen years later, Apollo XI landed on the moon. So it goes in science. The "blindingly obvious" way around the Perceptron's seeming limitations[9] lies in an only slightly more sophisticated understanding of XOR, one that by the 1960s had long been well known by logicians and early computer scientists who, of course, needed the function for standard programming. XOR *may be built up out of two of the three elementary logical operations,* AND *and* NOT. Here's how.

XOR means "Neither both nor neither." Now, "(NOT (A AND B))" means "not both," while "(NOT (NOT A AND NOT B))" means "not neither." Insist that both be true at the same time by putting an "AND" between them and you have XOR = (NOT (A AND B)) AND (NOT (NOT A AND NOT B)). This can be implemented with Perceptrons by stacking them in layers, so that it processes "in parallel," as illustrated in figure 3–2a.

The variable connection strengths are represented by ws; the θs are thresholds in each neuron that determine whether it will fire or not, depending on the net input. The problem boils down to this: Without telling it what to do, can we get this network to have its topmost neuron put out a signal ("1") when we feed it a "1" and a "0," or a "0" and a "1," but put out nothing ("0") if we send either two "1"s or two "0"s?

The answer is yes, so long as the pattern of weights eventually converges to one of a number of workable configurations. The fourteenth issue of the

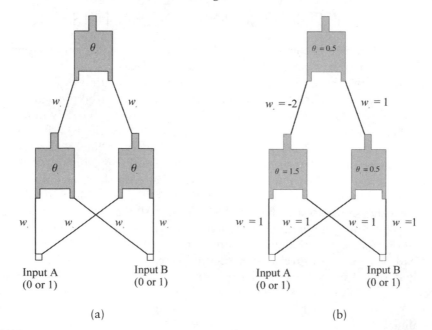

(a)   (b)

FIGURE 3–2   (a) Multilayer Perceptron ready for XOR problem. (b) Multilayer Perceptron after having solved the XOR problem

popular computer magazine *Byte* offered readers the solution shown in figure 3–2b.[10]

For example, suppose that the input is A = 0 and B = 1. Since this is a case of "one of the inputs equals 1, but not both of them," we expect the Perceptron to put out a 1. We take each of the two inputs, multiply them by the intervening weights, and distribute the result as inputs to the middle two neurons. Since A = 0, it will distribute nothing, so we can ignore it. With B = 1, and both the weights between B and the middle layer also being 1, both middle layer neurons receive $1 \times 1 = 1$ from it (and 0 from A).

Now, the leftmost middle neuron has a threshold $\theta = 1.5$. This means that if the total input it receives is 1.5 or greater, it will put out a 1; otherwise it will put out a 0. Its total input is 1 (from B) + 0 (from A) = 1, which is less than 1.5. So this neuron puts out a 0.

The rightmost middle neuron has a threshold $\theta = 0.5$. This means that if the total input it receives is 0.5 or greater, it will put out a 1; otherwise it will put out a 0. Its total input is 1 (from B) + 0 (from A) = 1, which is greater than 0.5. So this neuron puts out a 1.

The leftmost middle neuron output of 0 gets weighted by –2, and $0 \times –2 = 0$. This gets sent as input to the topmost neuron. The rightmost middle neuron output of 1 gets weighted by 1, and $1 \times 1 = 1$. This, too gets sent as input to the topmost neuron. The topmost neuron adds its two inputs together: $0 + 1 = 1$. It has a threshold $\theta = 0.5$. Since the total input is greater than this threshold, the topmost neuron puts out a final result 1, as expected.

If you want to get a hands-on feel for how neural networks of all kinds work, you might want to take a paper and pencil, feed in all the remaining possible inputs—(0,0), (1,0), (1,1)—and track the flow of numbers to see that the correct outputs are generated. Otherwise, you can take it on faith that it works—inputs (0,1) and (1,0) generate a 1; inputs (1,1) and (0,0) generate a 0.

So if part of the power of the Minsky and Papert critique of Perceptrons was in showing others what, in retrospect, seems an obvious problem—a typical characteristic of elegant mathematical expositions—part of its weakness was in its having missed the equally elegant, and equally obvious, solution to the problem.

PARALLEL VERSUS SERIAL COMPUTATION

There was a time, shortly after the Minsky and Papert critique, when the common wisdom prevailing in computing and scientific circles was that parallel computing was a dead end—it could do nothing "interesting," as Minsky and Papert put it. Well, it's worth making the following point: the truth is that *anything* you can do by programming a nonparallel ("serial") computer from the top down, a biologically modeled parallel computer can learn to do from the bottom up, just as Rosenblatt believed, and von Neumann before him. The reason is simple: *All computation consists of the elementary logical operations strung together*. Any system that can string these operations together, however awkwardly or inefficiently, can be a "universal computer" and compute anything computable.

It's not intuitively obvious (check it for yourself), but NOT XOR, for example, translates into English as "if and only if," as in "she's my wife if and only if I'm her husband," meaning "if one then the other *and vice versa.*" String a batch of NOT XOR nets together and you have a device that not only can *carry out* higher-order logical operations as instructed, it can search out the necessary and sufficient conditions for something to be true, on its own, and no matter how complex, whether the human designers of the net have discovered these conditions for themselves or not.*

So why haven't multilayered neural nets entirely replaced standard computation? For *already learned* processes that *can* be expressed in the form of an explicit rule, it is far more efficient to make direct use of that "expert" knowledge than to train an entire net to do it. This is precisely the trade-off that goes on in our own fourth brain: Its network functions learn a task; once learned, the completed algorithm for doing it is tacitly embedded. If you now make the rule explicit, you can depend on it in the future to guard against errors. That's how we acquire a specific language[11]: Training first, rules later. (In a later chapter we will see in more detail how a neural network acquires one aspect of language, natural pronunciation.) We learn to think critically not by learning the rules of "critical thinking" but by repeatedly exposing ourselves to situations where thinking is critical! Repeat *"Socrates is mortal"* often enough, and a bunch of other syllogisms, and you actually will begin to comprehend the rules of logic. Then when someone explains them to you, they'll seem "intuitively" obvious.

Standard versus network computation complement each other perfectly. Minksy and Papert's intuition that "multilayer" Perceptrons would prove sterile was just that: an intuition—and it was wrong. There was never a need for the proponents of one approach to attempt to do in the proponents of the other . . . right?

Wrong. The human antagonism endemic to science is itself a "competitive learning algorithm." At times it's painful to the individual "processing elements," but in the long run it benefits the larger network of humanity as a whole. As one scientist puts it, "Ruthlessness toward mistakes has always been one of the engines of scientific enterprise."[12] As far as I know, Marvin Minsky was never much bothered by being wrong—that's how one spontaneously learns, after all. Perhaps that's why he's so often been right otherwise.

---

*
  We haven't shown how a Perceptron converges on the solution to the XOR problem, or that it invariably must, only that there is a solution. Both the method and the proof exist, but they're beyond the scope of this book.

4

TEACHING A YOUNG DOG OLD TRICKS

This is the lesson: Never give in—never, never, never, never.
—Winston Churchill to the boys at Harrow,
October 29, 1941

Richard Feynman was typical of childhood prodigies. He found something he loved and never tired of doing it, again and again, an amazing number of times. This is *not* the high romantic definition of genius, nor does it sit well with the avant garde educational obsession to teach, directly, "critical thinking skills." It does comport nicely, however, with Dickens's famous definition of genius: "The infinite capacity for taking pains." So, by the time he reached high school, Feynman had puzzled his way through most of the problems in college calculus texts. By the time he finished college, he had repeated, over and over, and so had at his mental fingertips, thousands upon thousands of trials for almost every kind of mathematical and physical problem there was.

As a child, Feynman's younger sister, Joan, also wanted to pursue science. But their mother believed girls were physiologically incapable of learning science and strongly discouraged her, so she avoided it. But when Joan turned thirteen, Feynman gave her a college astronomy text. When she opened it, she saw that inside the cover he had written her name. She was deeply moved but concerned as she turned the pages.

"How can I read it? It's so hard."

His answer was the perfect description of how the fourth brain—an enormously powerful neural network—best learns, fearless of failure, undaunted by repetition, not seducible by educational fads and shortcuts: "You start at the beginning and you read as far as you can get until you are lost. Then you start at the beginning again, and you keep working through until you can understand the whole book." Joan received her Ph.D. in physics in 1958.

Try for yourself Feynman's advice to his sister, but keep a dispassionate tab on your own mental processes. As you repeatedly butt your head up against the same wall, you'll start to ask *"What am I missing?"* Mentally, or perhaps on paper, you'll toy around with various alternate interpretations of the sticking point. When you *stumble* upon the correct one, the roadblock vanishes and you proceed forward—until the next roadblock. If you track your progress, you'll see that you go forward in fits and starts. Significant forward movement follows a round of errors, *during which round blame for the error is properly assigned.*

Marvin Minsky and other neural network critics were so skeptical of even "multilayer" Perceptrons because he, and they, were intuitively convinced that *there is no way for a neural network properly to allocate "blame" for its errors to its various neurons*, especially if there was a large number of them in rows sandwiched between input and output.

SOMETIMES TWO WRONGS DO MAKE A RIGHT

Here's an example of such a multilayer network—a very simple example and one that's already been trained. That is, it's been presented with numerous apples and bananas and its connection weights have been *automatically adjusted* (how, we will get to shortly) in response. They are now just what they need to be for the network to correctly tell when it "sees" an apple and when a banana.

Instead of a neuron either being completely "on" (i.e., excited, black) and putting out a 1 or completely "off" (quiescent, white) and putting out a 0, these neurons are a little more lifelike and so have a range of possible excited states from black through gray to white. We see the network in two states: once when presented with a banana (left) and once with an apple (right). The connection strengths are now finalized so they stay the same in both cases. But which neurons get excited, and by how much, depends on whether the two bottom-layer input neurons sensitive to the features of bananas are stimulated, or whether the other two bottom-layer neurons sensitive to the features of apples are stimulated. Which of the three middle-layer neurons, and which of the two upper-layer neurons get stimulated, and by how much, also depends on the particular pattern of weights affecting the connections between all the neurons in one layer and all the neurons in the next. The middle-layer neurons are called "hidden" because they have no interaction with the "outside world," as do input and output layers. The output and hidden layer neurons are "computational" because they sum their inputs and apply a threshold value—or a set of threshold values—to determine what to put out. The input layer neurons are "sensory" but not computational because each gets excited in response to a unique feature presented to it (like specialized heat neurons and pressure neurons in our fingertips) and puts out a value that directly reflects how "intense" is the feature it detects. (See figure 4–1.)

If the network is presented with an apple (left), the "red" and "round" detector neurons in the bottom row light up; if with a banana (right), the "yellow" and the "long" ones do. The diagrams illustrate a pattern of weak connections (dashed lines) and strong connections (solid lines) such that the "apple" neuron in the top row gets more excited than the "banana" neuron when the network sees something roundish and reddish; and the reverse if presented with something longish and yellowish. The pattern of connection strengths "encodes" the detected features of apples and bananas. Taken together (super-

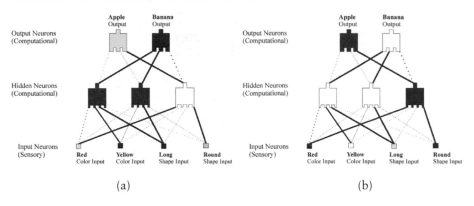

FIGURE 4-1   (a) Banana. (b) Apple.

posed, as it were) and distributed throughout the network as a whole, these encodings constitute the "image" or "memory" of the two objects.

It's not a hard problem for us, to be sure, but for decades most computer scientists were convinced that because of the scrambling of signals that takes place in the middle, "hidden" layer, and because these neurons cannot be given direct feedback (from us, for example), there *will be no way to have the connections automatically modify themselves to produce the above* (or an equivalent) *distribution of strong and weak links.* This encoding had to be done by an outside expert in fruits, claimed the popular wisdom.

Once the XOR problem had been solved, Frank Rosenblatt's Perceptron might properly be said to have earned its place in the Smithsonian. But without some relatively straightforward method of proving that in large, multilayer neural networks, the blame could *always* be properly assigned, the entire theory of neural networks was at risk of being hit by yet another impossibility proof, which, even if wrong, would set the field back yet another twenty years. After all, nobody was going to be interested in a computer that *might* give you the right answer.

In fact, by 1974, the problem had been solved—in a physicist's Ph.D. thesis[1]—and over the next decade or so many times over.[2] But researchers were then working in so many disparate disciplines—physics, engineering, pure mathematics, neuroscience, computation—that only in the 1980s did they become aware of each other's successes. In its final form, the answer has come to be known as the "error back-propagation algorithm," or simply *back-prop*, and it represented a crucial milestone in the development of biologically modeled computation.

The basic idea is this: The degree to which the network is wrong must be automatically *propagated backward through the network.* The person who builds the network doesn't decide how much to weaken each connection's strength but does formulate a fixed rule that says how severely the network is wrong (a fraction between 0 and 1). The network is built so that this fraction is sent backward through the network, weakening by a proportional amount those connections responsible for the erroneous outcome. Since different neu-

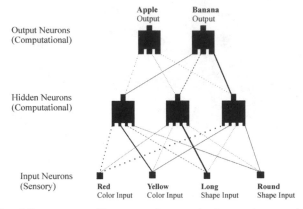

Output Neurons (Computational)

Hidden Neurons (Computational)

Input Neurons (Sensory)

FIGURE 4-2   Nine-neuron, 18-synapse (connection) neural network for solving the banana-apple discrimination problem.

rons get activated a different amount depending on both the connection strengths and different inputs, strong connections between very active neurons that generate bad answers will bear the brunt of the blame and will be "punished" accordingly. It's like a general who having lost a battle distributes everywhere the message: "That was *terrible.*" Officers in battalions responsible for the battle take the blame. If "no promotions, two demotions per battalion, for a terrible loss" is the rule, then no promotions and two demotions quickly happen in that battalion. Battalions not involved in the fiasco are untouched.

By giving such a network thousands of examples to chew on, over and over, it will learn to detect patterns in the data that might remain invisible even to a sophisticated human expert. Consider this: Figure 4–2 shows the banana-apple network before it's been trained to discriminate between apples and bananas.

At the start, all neurons in the network are quiescent (black) and the connection strengths are random but near zero (densely dashed lines a bit below zero, solid but thin ones a bit above). We present the network with a training set of, say, 50 each bananas and apples. We'll be kind to our baby network and avoid a large proportion of obviously confusing fruit, such as yellow apples and truly reddish bananas—just as we are careful in teaching the basics to young children.

Suppose the very first thing we show it is an apple. The "red" and "round" neurons light up—in figure 4–3a, we're catching it before they've had a chance to pass on their excitation to the middle, "hidden" neurons.

This causes the hidden, and then the output, neurons to become active to varying degrees, depending on the (random) connection strengths. (See figure 4–3b.)

Because of, for example, the strong connection between the leftmost hidden neuron and the rightmost output neuron, the network puts out "Banana" more strongly than "Apple." The exact degree of error—just how wrongly excited (lit up) is the neuron—now gets propagated back through the network,

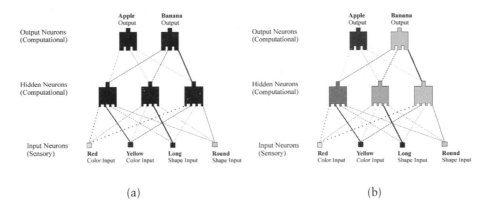

FIGURE 4–3   (a) An apple has been presented to the network. Therefore, the "red" and "round" feature-detecting neurons (inputs) are active. The excitations of these neurons have not yet reached the hidden and output neurons. (b) The "red" and "round" feature-detecting neurons (inputs) excite the hidden neurons, and these in turn excite the output neurons. The level of activation is a function of (a) how active the preceding neurons are and (b) how strong the connection strengths are.

and the connection strengths are adjusted. Eventually the network converges on a distribution of connection strengths (there may be many) like the one above that correctly distinguishes between apples and bananas.

THE HOLOGRAPHIC BRAIN CATALOG

Without ever having been taught that "bananas are long and yellow, apples are red and round," the network has "deduced" this rule on its own. Deduced is in quotes because, really, the rule is nowhere made explicit, it is only *tacitly* embedded in the final pattern of connection strengths (weights). The *images* of banana and apples have been *memorized* by the network—in just the same way (or very similar to the way) that we memorize images of things so as to be able properly to categorize them.

Where are the images of banana and apple stored? Nowhere and everywhere. Each is *distributed in parallel over the network as a whole*—hence the term "parallel distributed processing," which obscures the biological, and human, origins of the method. Both images "occupy" the same network of synapses, and neither can be said to be located in a single or unique location.

As the network memorizes the difference between apples and bananas, it makes fewer and fewer errors—on the whole. As did HER, it sometimes slips backward. But the more it slips, the stronger the error message, the faster it once again leaps forward. *Recueillez pour mieux sortir*, as the French say: "Fall back, the better to advance." The average error over time looks like figure 4–4.

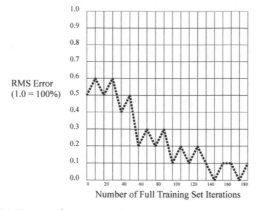

FIGURE 4-4   RMS error during training.

PICKING WINNERS

Every city of any decent size in America has at least one stock broker's office. On any workday, one may wander in and witness a scene that has changed little for well over a century: Seated in a row of chairs are a group of retired men. They sit for hours, chatting about their lives and their children and their grandchildren. As they talk and laugh, their eyes are constantly glancing at a swiftly moving line of electronic letters and numbers.

These are America's "tape readers"—amateurs who have invested their retirement nest eggs in stocks and who make their decisions based not on "portfolio theory," or "price-earnings ratios," or "book-to-market value," but on hours and days of fleeting symbols streaming into the retinal "input layers" of the fourth brain.

The input process rarely disturbs their amiable storytelling. Long ago each man assigned to a subset of neurons the task of taking in the data and of processing it. The men "read" stock prices the way that we drive cars—thoughtlessly, expertly, half unaware, our "minds" elsewhere.

On the whole, these little players are pretty savvy: They rarely get rich (and are wise enough not to try), but they often turn barely livable pensions into quite comfortable ones. How do they do it?

Here's a quick glance at a scaled-down version of actual trading networks now in use at every major investment house, networks that mimic the experience-based, theoryless "intuition" of our line of "players." The network is composed of 28 neurons: 20 in the input layer, 7 hidden neurons, and just 1 output neuron. All neurons in the input layer connect up directly to the output layer as well as to the hidden layer. Thus *all* layers are interconnected with one another. The 28 neurons are thus connected by a total of 167 modifiable weights.

Every evening we feed the S&P 500 closing value for twenty prior trading days into its 20 inputs. The output neuron puts out the network's prediction of how much the next day's closing value will differ from the last input day. We teach the network *nothing* about investment strategy, economics, or business. We simply tell it an old twenty-day sequence of S&P 500 closings, and let it guess what the next day's value will be. We then compare its guess to the real value and feed the error back to it. It responds to the error by altering weights according to the back-propagation algorithm. Then we give it another sequence and then the new error. This we do over and over until it's absorbed whatever is going on that underlies the changing value of the index.

More exactly, we're going to feed it 760 consecutive trading days' worth of S&P 500 closings, from August 4, 1989, until July 2, 1992—that's 760 sets of twenty days of data = 15,200 total "inputs" per set. We will then feed the complete data set to the network 50,000 times. The network will have therefore absorbed and processed, with no instruction, just feedback as to how it's doing, a total of 760 million pieces of data. Because of the speed of modern computers, training of this sort requires all of 40 seconds—somewhat more rapid than the brains of our retirees.

To give you a feel for the complexity of such a network's interconnectedness—and for the staggering amount of signaling taking place—figure 4–5 is a schematic of what the network looks like.

After training is complete, the weights will have settled into some final pattern. That's because if the network "converges," the error grows smaller and so, too, the amount of adjustment to any weight. When any further weight changes are insignificant, we turn the network loose and allow it to predict the twenty-first day's close following 20 days of input. We do this for a total of 240 days: starting 20 days following the last training data (here July 22, 1992) and ending 240 trading days later (on June 23, 1993).

Note that *the network must base its predictions on data it has never seen before*, based solely on its prior experience. If we only train it once—and don't force it into "continuing education"—the training data is going to grow ever more dated. If the relationships "intuited" by the network between past and future performance slowly change over time, its prediction will grow less accurate.

Figure 4–6 shows the results. First, we show a standard measure used by traders to guess at market trends, a so-called moving average, and compare it to the actual closing prices for the S&P 500 for our 240-day test period. The moving average consists simply of the numerical average of the preceding 20

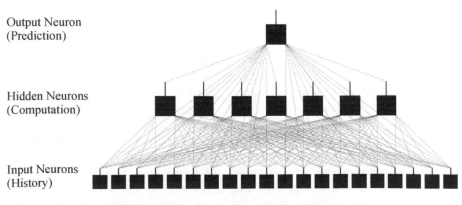

FIGURE 4–5   S&P 500 stock price prediction network.

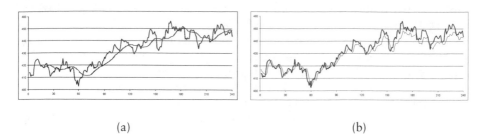

(a)   (b)

FIGURE 4-6   (a) S&P 500 stock index (heavy line) versus 20-day moving average.
(b) S&P 500 stock index (heavy line) versus network prediction (light).

days used as a prediction for day 21. This is compared to the network prediction compared to the actual data.

The network is clearly superior to the moving average, especially early
on. After about 90 days, however, the network prediction begins to diverge no-
ticeably from the actual market index. Whatever tacit rules our network has
come to embody in the pattern of its weights are evidently not fixed—they
change over time, leaving our network behind.

How can the network adapt to these changes in the hidden rules governing stock prices? In the same way our retired investors do: They do not spend a
few years training on the ticker and then apply what they absorb without further training. On the contrary, after an initial period (perhaps) of investing on
paper, they continue to learn from their mistakes. So it is with the actual artificial networks now being used by investors to supplement the biological ones
inside their heads. Every evening, after the markets close, the network is retrained on (say) four years of preceding data, *excluding one day at the beginning, and adding the current day's data.*

The typical network used for market (and other financial) forecasting
will chew on at least 100 days' worth of closings. Why not 1,000? It's been
found that at least five training data points are desirable per connection, and in
a network of 100 input neurons there are over 5,000 connections. To avoid
making spurious inferences (the hallmark of low intelligence!), 1,000-day trading network should be trained on at least 12,000 *accurate* market closings.
That's about fifty years' worth of high-quality information, which is not so
easy to come by.

Assuming it may be had, that's about 60 *billion* pieces of information
that the network must process freshly each night if it is to be properly retrained
before the market opens the next morning, and it must cycle that data through
more than thirty times the number of connections as in our model stock predic-
tor above.

Just a few years back, only the largest securities firms in the United States
and Japan had the available resources to do this sort of computing (as did a
few very savvy individual investors with the foresight to buy older supercomputers, which plummet in value as soon as the next model is available). Today
a 700 MHz computer can do the job in about 20 hours—good enough for the

weekend, maybe, but otherwise it's too slow. Therefore in both the United States and Japan, large-scale projects are under way to create huge *hardware-based* neural networks, in which the "neurons" are not modeled in software but burned into special chips themselves. These neural chips will be able to do the job at least 1,000 times faster than a standard supercomputer.

As an example of what can be done using desktop neural nets implemented in software, consider one such net that was used to trade commodities on a daily basis. After nine months it returned more than $103 for every $1 invested[3]—a better track record than even the infamous Arkansas network of connections.

NETTALK

One of the great landmarks in the development of practical neural network technology *and* in our understanding of the brain-as-neural-network was a 1987 speech synthesizer called "NetTalk."[4] NetTalk was the first computerized device that could read written English text and pronounce it correctly. This is an extraordinarily difficult task. Even the *formal* rules of English pronunciation, as written out at the beginning of a dictionary, are highly inconsistent, with many exceptions. The actual, *tacit* rules—however instantly the native English speaker reacts to their violation—are far too complex ever to have been made explicit. Standard computers never got anywhere with speech synthesis, because there are no "experts" who can do an even halfway decent job of codifying the rules. Yet within a few years almost every infant learns them perfectly.

The challenge in speech synthesis is that the correct sound of a letter or letter-combination depends in so complicated a way on the letters that precede and follow it.* Consider the sentence, "This chair is a chaise." The s in "th*i*s" and "*is*" are pronounced differently yet both belong to the two-letter sequence "*is*." The *ch* in *ch*air and *ch*aise are pronounced differently, even though both are the first two letters of the same four-letter sequence *chai*. Likewise, the last two letters in *chai* are pronounced differently in ch*ai*r and ch*ai*se. By the sixth grade, a reasonably well-educated elementary school student should have no difficulty pronouncing more complex examples: "Sherry, this cherry-colored chair is a chaise." But for decades, computerized speech-synthesizing programs were stiff and weird-sounding. They could hope to sound at best recognizable, not human. Then came Terence Sejnowski.

After graduating from Case Western at the top of his class, Sejnowski went to Princeton. There he fell under the spell of neural networks—while he

---

* Some pronunciations also depend on meaning, but researchers would soon reason that meaning itself correlates almost always with the immediately preceding and following string of letters—taken as *whole*. Exceptions involve long strings with multiple heteronyms, such as read, sow, bow, and tear. For instance, in the sentence, "You know, the lead is a really big problem," you can't tell whether it's "the *led*," a deadly metal, or "the *leed*," a deadly actor. You need a context larger than the immediately preceding or succeeding letters.

was supposed to be studying for his qualifying exams in physics. This was not so odd a departure as it may seem, for theoretical physics and neural network theory turn out to be closely related.

In 1985, just before leaving Princeton, Sejnowski gave a talk that fired the imagination of Charles Rosenberg, a graduate student in psychology who was also a crack programmer. Rosenberg approached Sejnowski looking for a summer project, preferably something to do with language. Sejnowski was intrigued and decided to take it on. Rosenberg put together a back-prop network pretty quickly. He programmed into it no rules of pronunciation or of spelling. Its "method" was exactly that of children: It made many silly mistakes, repeated, made fewer and less silly mistakes.

The designers of NetTalk adopted the most typical set of phonemes found in American English. The input layer network was fed a string of text including punctuation and spaces. It then put out one final signal per letter, which was fed into a sound generator. Now, the crucial clues as how precisely to pronounce a letter lie within the context—the letter by itself is nowhere sufficient. So, Rosenberg and Sejnowski took successive "windows" of seven letters as the basic input, sliding the window along the string of text: three preceding letters, three succeeding, and the letter itself.

At the beginning, the sounds the net generated for the letters it was "reading" were selected completely at random. Sejnowski and Rosenberg had devised a rule that assigned a value according to how badly the network mispronounced the middle letters. For each sequence of seven letters the error value was automatically calculated and propagated back through the connections. (The network was never told what the correct pronunciation is.) These connections were modified accordingly, and the network slowly fumbled its way toward ever better pronunciation.

How big was NetTalk? For each successive letter in the text, the net received a total of seven letters: the letter itself, three letters before, and three letters after. Given the twenty-six letters of the English alphabet, plus space, comma and period, all times seven, the network used a total of $29 \times 7 = 203$ input neurons. With each presentation of a letter, seven of these are activated.

The hidden layer consisted of 80 neurons,* and the output layer of 26. Twenty-three of the output neurons were dedicated to the generation of individual sounds; three were dedicated to syllabic stresses. The net output signal consisted of various *combinations* of these twenty-six neurons' individual outputs. The single most heavily weighted output signal was fed into a digital converter to produce an audible sound.

The network had roughly 4,000 connections to be varied by training. Even this network is extremely simple compared to those actually at work in the fourth brain. Figure 4–7, therefore, will give you an even better sense of the

---

* The optimal number of neurons for the hidden layer in a back-propagation network is itself usually determined by trial and error and by certain general rules of thumb that have become important over the years. But often, the number is dictated by how much reliable training data is available and how much can be processed using available resources in the available time.

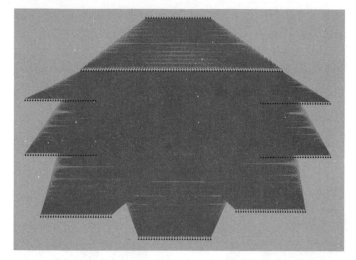

FIGURE 4–7 Schematic of NetTalk showing neurons as tiny black squares with small vertical projections representing axons (as in prior schematics, but at a much smaller scale). *All* the connections have been drawn in using thin gray individual dashed lines. The resulting solid color is due to the large number of these lines, up to 203 per input neuron. (The input neurons have been segregated into seven groups to enhance the visibility of their connections.) Compare to the human fourth brain, wherein each neuron may both send and receive as many as 10,000 connections.

extraordinary complexity of the connections that underlie human behavior and intelligence.

NetTalk successfully learned to pronounce English text, as expected. But the most striking feature of its performance, noted by many observers since, was the *way* it sounded as it progressed: *the same way children do.* "The network started out by babbling . . . " writes one author, but "the distinction between vowels and consonants was made early. A second stage occurred when word boundaries were recognized. At that point, the output sounded like pseudo-words,"[5] a stage recognizable to any parent of an infant just learning to speak, who lies happily in his crib "pretending" to talk.

After being presented 50,000 times with a selection of 1,024 words, Net-Talk was correct 95 percent of the time when tested on that training set. When presented with words it had never before seen, its pronunciation was 80 percent correct. Compare this to the schoolchild who reads a new word—how often does she pronounce it properly, needing no further correction?

Nonetheless, as complex and capable as artificial neural networks have become, that complexity pales by contrast to the human fourth brain on which they are modeled. It is to that we now turn.

5

THE WET NET

Mind: A mysterious form of matter secreted by the brain. Its chief activity consists in the endeavor to ascertain its own nature, the futility of the attempt being the due to the fact that it has nothing but itself to know itself with.
—Ambrose Bierce, *The Devil's Dictionary* (1906)

In 1913 a 54-year-old Spanish histologist from Petilla de Aragón, Spain, Santiago Ramón y Cajal, was awarded the Nobel Prize in Medicine for his "neuron doctrine"—the hypothesis that the nerve cell is the fundamental unit of the nervous system. Figure 5–1 is one of his now-famous drawings of a thin section of brain tissue, stained with silver to highlight the neurons.

Though the section has the typically organic and tangled appearance of biological tissue, there is a great deal of evident order. Nineteen pyramidal neurons are arranged roughly in two layers, the first consisting of the eight larger cells, the second of eleven somewhat smaller ones. All appear oriented with a vertex pointing upward, and out of these vertices project, also vertically, long thick strands that, after a while, branch horizontally. Into (or out of) base corners of the pyramids there extends a great many somewhat more finely branch-

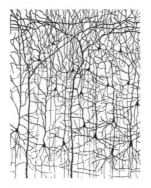

FIGURE 5–1   Silver stain of human visual cortex drawn by Ramón y Cajal in the early 1900s.

ing projections. It's the same structure employed in artificial neural networks: one in from each; one out to each, as in our banana-apple network.

But Cajal's painstaking figure is of an almost unbelievably tiny fraction of the human cortex—it captures less than *one-billionth* of the number of cells in the fourth brain. Does it really seem possible that 20 billion neurons could be wired together according to a fixed scheme, as neuroscientists (until recently) have long insisted? It is beyond possibility once we realize that Cajal's stains not only eliminated surrounding nonneuron tissue (his great histologic accomplishment), they also eliminated the vast majority of connections between neurons. In a living brain, each cell has *1,000 times more connections.* Cajal's drawing nicely (and accurately, as it turns out) captures the general *pattern* of connectedness among neurons. But it utterly distorts the actual *density* of connections in the living brain. These connections form an almost inconceivably dense three-dimensional web of links among neurons.

In the human fourth brain, *every neuron sends out and receives up to 10,000 connections.* Only so long as this level of complexity remained unsuspected was it possible to think of the brain as wired liked a radio or a conventional computer.

A handful of artificial neurons with a dozen-plus connections can learn to tell the difference between apples and bananas. A dozen with a few score connections can be taught to play Hexapawn or tic-tac-toe, perfectly, in a few seconds. Fifty or so, with about 1200 connections, have in reality done a better job of diagnosing heart attacks in an emergency room than a team of expert cardiologists. A hundred neurons, with a few thousand connections, can make money—a lot of money—in notoriously volatile markets. A few hundred can mimic the development of human speech. Of what, then, is the 20 billion neuron, multiquadrillion connection human brain capable? No, let's put the question differently: Is anything of which we already *know* the human brain capable *not* within the reach of a large enough, completely mechanical, neural network? We can answer that question only after first considering additional subtleties of the brain that have been successfully incorporated into artificial networks to make them work even better.

YES, GENTLY INTO THE NIGHT

Today, our lives are filled with miniature computing chips. They regulate the speed of our automobiles, the temperature in office buildings, the alarm systems that guard our property—and too many thousand other examples to list. The vast majority of these chips use standard rule-based programming "hardwired" into the chips themselves. They are made of hundreds or even thousands of tiny subunits wired together according to a precise plan. If even one of those subunits should fail, or if even one of the tiny gold wires inscribed on a circuit board should break, the entire device fails.

One of the puzzles in the early days of neuroscience was the so-called fault tolerance of the brain. Early researchers well knew from the extreme cases—brain trauma, infections, degenerative diseases (i.e., Alzheimer's)—that *millions* of neurons can die, and therefore untold *billions* of connections can be

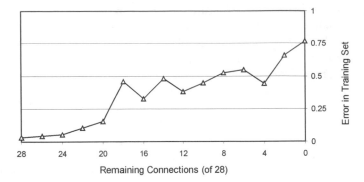

FIGURE 5–2   Root mean square error of a network, but with different numbers of connections randomly selected from among twenty-eight possible ones.

severed, without any seeming effect. Given the biological nature of the brain, many thousands of neurons must die in the normal course of events without the slightest detectable loss of function. How can this be? The answer is simple, with a practical elegance that is the hallmark of nature's solutions to challenges: The brain uses randomness, and a computer that does *that* doesn't mind much if its innards get scrambled a bit—they're scrambled to begin with!

Here's how it works: So far, we've described networks with *every* allowable connection between layers of neurons. What would happen were we randomly to eliminate some of them? That's the same thing as wiring a net up from the beginning with not only an initially randomized distribution of connection strengths ("initialized," one says), but with a randomized set of connections, too. Here's an example. We train a twenty-eight-connection network (this one classifies flowers), note the final error, randomly destroy two of its connections, reinitialize it, and repeat the process until no connections remain. Figure 5–2 shows each of these trials, with the little triangle showing how often the network is wrong after training has been completed.

We see that not until there are roughly 25 percent fewer connections does the network begin to shown significant signs of dullness—as indicated by a decreased ability to learn. Furthermore, we see that it loses IQ not all at once, as would happen with a preprogrammed circuit, but gradually—"graceful degradation," in the lingo of computer and neuroscience. Stretch the horizontal axis out over not a mere twenty-eight connections, but over 10^{14} of them, and you can understand why the fourth brain is so resilient. The loss of neurons proper yields similar results; you may, if you wish, think of the loss of a neuron as equivalent to the loss of all its connections.

We can use this to understand the insidious onset and typical course of degenerative diseases such as Alzheimer's disease. Alzheimer's gets silently under way for many years before the loss of neurons becomes large enough to cause noticeable cognitive degradation. By then the loss of tissue is so great that it can be seen on the relatively coarse scale of brain computed tomography (CT) or magnetic resonance imaging (MRI) scans.

If we look at the same phenomenon from a slightly different point of view, we can understand another perplexing feature of Alzheimer's disease: the loss of so-called short-term memory, which is really another way of describing *a decreased* (in the extreme, wholly lost) *ability to learn.* (To see their equivalence, think of some of the gallows humor that circulates in professional and caregiving circles responsible for people with degenerative dementias: "You know, Alzheimer's is not so bad. Think of all the new people you're always meeting.")

But, there's another side to this. Alzheimer's is generally a disease of old age. The neurons (and connections) lost to it are ones that take part in *an already "converged" network,* converged onto the set of solutions, responses, and memories that constitute the unique person. Many of the connections that existed in youth would have weakened anyway, as part of a lifetime of learning from experience. But others have grown strong and play a critical role. Their loss is devastating. But the loss of connections *before* convergence—in youth, that is—might not be so bad. In fact, it turns out that the carefully timed *pruning* of select connections actually hastens convergence toward zero error. Only if the pruning is excessive or ill-timed does it hamper or destroy the intelligence of the network, as we see in figure 5–3.

Pruning is in fact a normal process in the fourth brain that accelerates greatly during adolescence. Why it happens is easy to understand: Learning is both time and energy consuming. A reduction in the number of connections translates into potentially significant "cost savings"—central processing unit (CPU) time for the calculations in artificial networks; energy available to the brain for all its tasks (the human brain as a whole expends a remarkable 25 percent of our energy intake while consisting of only 2 percent of our living body tissue mass).

How pruning happens once again reveals the subtlety of biological solutions. We know from Alzheimer's disease that if established connections are to be *randomly* reduced, it better not happen *after* training. But what if there were a way of identifying relatively *unnecessary* connections, the ones that have become weakened over time? These could be eliminated without risk. At first blush it would seem that this would require some kind of outside intervention. But, in fact, all that is required is that weak connections become physically frail, as well. They will then be vulnerable to general influences—alcohol, trauma, infections, fevers, hormonal fluxes—that have no effect on stronger connections. This is just what happens, especially under the rush of hormones during adolescence. In biological systems (by contrast to mathematical models) there's no need for thoughtful erasure—the weak perish on their own. "Connection strength" in living neurons is not a mere number or a line of code, it is a measure of the actual physical density and robustness of certain tissue structures—synapses.

As always, however, there is no free lunch: Pruning a network tends to "fixate" an already adapted network, whether natural or artificial, on the specific problem and data on which it has been trained. If we now ask it to change—to adapt to a new set of data, or to learn some new categories on top

The Effects of Pruning: Normal and Pathological

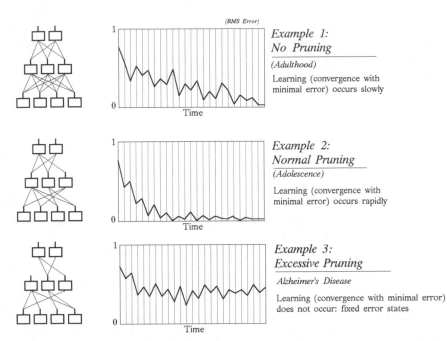

FIGURE 5–3   Pruning is a two-edged sword. It has beneficial effects on the speed of learning new but related tasks. If it goes too far, the network learns poorly at best, getting worse as the pruning continues.

of the old ones—it will be far less able to do so than a network with a full complement of connections. After adolescence, our brains are never again quite so flexible as when we were children. "Youth may indeed be wasted on the young," yet "there's no fool like an old fool."

SO, JUST HOW DO YOU TEACH
AN OLD DOG NEW TRICKS?

Does this mean that it's impossible to learn something new while *unlearning* something old? Well, there's a good deal of everyday experience that tells us just that. Infants absorb the nuances of language at an absolutely astounding rate—and to a degree of perfection that can never again quite be matched upon learning a second language later in life; schoolchildren absorb wholly new subjects at an almost equally breakneck pace, learning the habits and essentials of an entire civilization in about a decade; older people who return to schooling after a long absence find the adjustment notoriously difficult; physicists, mathematicians, and musicians are commonly reputed do their best work in their twenties or thirties and then to coast, or even decline, thereafter.

Indeed, the entire premise of the psychotherapy industry is that patterns of thought, feeling, and behavior are learned in childhood. To alter them requires a massive cash investment to achieve a rather mysterious "transformation." (The managed care industry disagrees, of course—for similarly pecuniary reasons.)

There is a widespread impression, usually tacit, that "change" means getting wholly rid of old patterns and replacing them wholly with new ones. Anything short of total replacement is taken to mean one of two opposite things: either "the condition is therefore hopeless" or "the condition is therefore normal," depending on the fashion and financial incentives of the day.

The operation of neural networks sheds light on these mistaken notions and may have some possibly discomfiting implications regarding what we can expect from ourselves and others. In brief, you certainly *can* teach an old dog new tricks, but it is more difficult to do so, depending on what his old tricks are, and whether he is a dog willing to pick himself up and start all over, again and again, or not. Psychotherapeutic change is fundamentally little different from mere learning (*pace* Freud, Jung, et al.)—chiefly because it shares with learning the same kind of obstacles: laziness, the quest for novelty, the desire for instant gratification, and especially the avoidance of failure. We don't want to remember the famous men who had to fall to rise again, because it suggests that to meet our own challenges we might need to do the same.

The solution to the old-dog problem is practical and pedestrian: repetition, perseverance, courage, willingness to start all over again. New patterns may be embedded in a fresh network relatively quickly. When a network is already converged to a specific set of patterns and partially pruned, it will require many more repetitions to "overlearn" a new—especially a contradictory—pattern.

Furthermore, just as a network stores images like a palimpsest, "one on top of another," it stores lately added patterns the same way. If the new pattern is repeated deliberately, diligently and frequently; and if the old pattern is willfully avoided for a long time (however natural the old pattern feels—*second* nature, not *first*, is the apt description), then even very old habits of thought, feeling, and action will slowly weaken as the newer ones grow strong. To think that old ways will ever disappear entirely—or to insist that, for a "cure" to be genuine, they must—establishes expectations certain to be disappointed. Because, given how the brain really works, "the perfect" is truly "the enemy of the good." To continually improve, one must continually repeat, but to be willing continually to repeat, one must be willing to fail, again and again. Behind many a fancy educational and therapeutic fad lies a hidden, all-too-human vice: the fierce desire to avoid failure, at any cost. That's the real reason that "rote memorization" and "drill" have acquired so bad a rep these days. Yet these mundane chores are precisely what turns the fourth brain from a mass of randomness into a intellect of dazzling capacity. "Genius," according to Thomas Edison, "is one percent inspiration and ninety-nine percent perspiration. Of "critical thinking skills," he had nothing to say.

GOING IT ALONE

Nonetheless, well into the 1970s, the idea that the brain might be a fantastic-ally complex neural network was widely thought ridiculous in academic cir-cles. For even if back-propagating neural networks could compute, the human-brain-as-neural-net hypothesis required a giant leap over a gaping hole: *Every such neural net requires external direction.* The Hexapawn robot, HER, re-quires *us* to determine what constitutes a win and what a loss; it's *our* under-standing of the rules of the game that allow it to learn. Stock market nets, text-to-speech nets—they all require someone outside the net to decide what constitutes the net's error and then to design the thing so that this error is prop-erly propagated *backward* through the very same connections that do the pro-cessing. No living neuron had ever been found even remotely capable of such a thing.

In short, back-prop networks learn "by themselves" only in the fashion of a child playing "hot and cold." You know where the prize is hidden and your daughter doesn't. So what if you don't hand her a sheet of paper with an X marking the spot where the treasure is hidden? The sequence of tempera-tures you call out in response to her initially random probing (cool, cold, warmer, hotter, really hot! . . .) is in fact such a map—just coded differently. She doesn't *really* figure it out on her own. It only seems that way to her, because she's a child. Her learning, like the back-prop algorithm, is "super-vised."

The eerie relationship that learning machines bear to human learning ac-tually has very little to do with how the brain can realistically operate at the detailed scale of its web of neurons. Perceptrons, multilayer Perceptrons, and complex neural networks are all models of high-level trial-and-error learning of a pretty mundane sort.[1] But *within* the brain, individual neurons networked within neural tissue do *not* get direct feedback as to how they're doing. Yet a vast proportion of what the brain learns to do via trial and error—to direct and coordinate crawling, for example—it does without any kind of supervi-sion. The brain, in other words, is able truly to learn spontaneously, yet re-quires no fancy method, it seems, to deal with error. Neural nets might *look* like how the brain is wired, but they are obviously completely different.

Between the attack on nets from the computer side and their obvious flaws as true biological models, it took a very convinced and very courageous individual to hang tough with neural nets during the 1970s, when both com-puter science and brain research were exploding. This was the era when the groundwork of today's great information-processing and biotech fortunes was being laid. If you were a fan of neural nets, ". . . it went roughly this way: you start telling somebody about your work, and this visitor or whoever you talk to says, 'Don't you know that this area is dead?'"[2]

The few who stuck with it were neither computer scientists nor neuroscientists, they were mostly mathematical physicists, who, like John von Neumann before them, were gripped by a deep, intuitive conviction that, how-ever unlikely it seemed at present, there lay within neural network theory some great secret of intelligence—perhaps even of life itself. They also knew that for neural nets to be taken seriously, something wholly new was going to be needed: *a neural network that could organize itself without supervision.* And its neurons would somehow have to accomplish this without some jerry-rigged system of feedback that couldn't possibly exist in nature. If it could be done, the payoff would be doubled: Neural nets would stand fully vindicated as a new method of "programming" computers—they would, in effect, program themselves; neural nets would stand vindicated as an explanation as to how the brain actually works, by *organizing itself.* This was a very tough lock to pick, but the key had actually been lying around for half a century.

COMPUTATIONAL MAPS

Cajal's careful drawings of brain tissue (see figure 5–1) had almost become a logo for the field of neuroscience and for the mysteries of brain circuitry. Following in his footsteps, in the late 1930s, another great Spanish neuroanatomist, Lorente de Nó, began to search for a pattern in the jumble of connections. He found one. He showed that, in fact, throughout the fourth brain, a typical "wiring" pattern appears again and again among clusters of neurons. The pattern always varied slightly, and at random, it seemed, but it was always recognizable and it was everywhere. It came to be known as the canonical cortical microcircuit. Figure 5–4 shows it progressively extracted from the jumble of cell bodies, axons, and dendrites revealed by Cajal's silver stains.

Lorente de Nó's finding tantalized neuroscientists for five decades. The specifics of the circuit he worked out readily enough, but no one could figure out what it was supposed to do. Its main features were wholly unexpected in an organ thought variously to work like a hydraulic system of pipes and valves, like an electronic signal processor, or like a logic-based computer. For within a microcircuit, *any two neurons send and receive connections to and from each other* (i.e., out from neuron 1 and in to neuron 2, and also out from

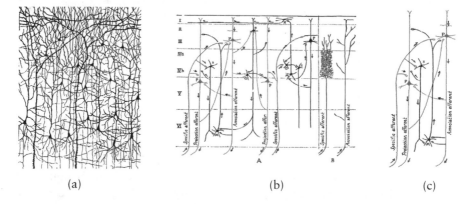

(a) (b) (c)

FIGURE 5–4   Microanatomy of the cerebral cortex. (a) Cajal's sketch from the early 1900s. (b) Sketch by neuroanatomist Lorente de Nó, ca. 1930, "to exemplify the broad plan upon which the central nervous system is organized," deduced from (a). (c) Single "canonical microcircuit" extracted by the author from within the middle sketch.

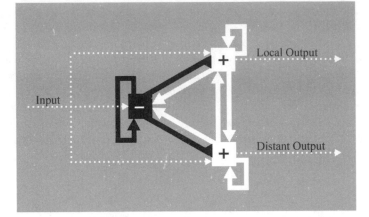

FIGURE 5-5   "Canonical microcircuit" of the cerebral cortex in schematic form.

neuron 2 and in to neuron 1), and often these connections inhibit the firing of
one or both. Like the microphone/loudspeaker feedback that plagues amplifiers, circuits with those kinds of loops are apt to overload when the reciprocal
connections are excitatory; but they're likely to shut down altogether when
the reciprocal connections are inhibitory. If the mutual connections are mixed,
the system could end up oscillating like the knocking of pipes in an old steam-
heated apartment building. It was later discovered that neurons usually send
out both excitatory and inhibitory connections to themselves. There was no
model that could make sense of such a design, shown schematically in figure
5-5.

Working in almost complete isolation between 1972 and 1982, a Finnish
physicist–turned–computer engineer named Teuvo Kohonen incorporated precisely these features into a two-dimensional sheet of artificial neurons, allowing connections to run from any one neuron to a surround of neighbors. The
results were amazing: He discovered that the pattern of connection strengths
between such a sheet of neurons would indeed *self-organize, without any supervision.* The sheet would spontaneously form a "map" of whatever pattern
of data was presented to it. Limited only by the total number of neurons in the
sheet, a great many different such maps could be embedded—*memorized,* as it
were—by the sheet. It was a tour de force demonstration of almost exactly
how the brain must do likewise. Today Kohonen's pattern-detecting Self-Organized Feature Maps are a crucial component of military aviation and subma-
rine detection devices. At the time, however, they were almost completely ig-
nored.

If nothing else, Kohonen's maps highlighted just those features that make
unsupervised self-organization possible: very large numbers of more or less
identical elements with both excitatory and inhibitory reciprocal connections.
The key was establishing oscillations, and that's just what the canonical microcircuit does, exceedingly well. Local clusters of processing elements set up stable oscillations, and such a cluster therefore acts like a larger processing ele-

ment itself, in mutual interaction with other clusters. Up the scale it goes—until the entire net establishes some kind of stable state.[3]

Data in a fully self-organizing system therefore does not flow in just one direction, it sloshes back and forth (along different channels, however); or, if there are many elements, it "circulates" in complex patterns. If the sloshing or circulating settles into some kind of more or less stably repeating pattern, it is called "oscillatory."

Teuvo Kohonen's self-organizing maps work, impressively. But they never caught the full attention of the scientific world. The reason is simple—it's impossible to understand exactly *why* his maps work, in the mathematically certain way that Minsky showed that single-layer Perceptrons couldn't possibly work (the same way that others showed that *back-propagation* networks must work). A mathematical description of the behavior of artificial self-organizing maps lies even beyond even our present abilities.[4] Neural nets continued to languish, therefore, and their intimate connection to the human brain was doubted until 1982, when a single event brought neural network research out of exile, established the new field of computational neuroscience, and sparked an explosion of development and intellectual ferment. This event was the presentation of a simple, perfectly describable model of spontaneous memory formation. Once again, the feat was accomplished by a physicist. It brought brain research one step closer to the natural, subatomic domain of physics proper.

6

SPINNING THE GLASS FANTASTIC

She sent for one of those squat, plump little cakes called "petites madeleines." I raised to my lips a spoonful of the tea in which I had soaked a morsel of the cake. No sooner had the warm liquid mixed with the crumbs touched my palate than a shudder ran through me and I stopped.

What could it have been, this unremembered state. I feel something start within me, something that leaves its resting-place and attempts to rise, something that has been embedded like an anchor at a great depth; I do not know yet what it is, but I can feel it mounting slowly; I can measure the resistance, I can hear the echo of great spaces traversed.

And suddenly the memory revealed itself. The taste was that of the little piece of madeleine which on Sunday mornings my aunt Leonie used to give me, dipping it first in her own cup of tea. Taste and smell alone, more fragile but more enduring, more unsubstantial, more persistent, more faithful, bear unflinchingly, in the tiny and almost impalpable drop of their essence, the vast structure of recollection.

Immediately the old grey house upon the street, where her room was, rose up like a stage; and with the house the town, the Square, the country roads we took. In that moment all the flowers in our garden and in M. Swann's park, and the water-lilies on the Vivonne and the good folk of the village and their little dwellings and the parish church and the whole of Combray and its surroundings, taking shape and solidity, sprang into being, town and gardens alike, from my cup of tea.

—Marcel Proust, *Remembrance of Things Past* [condensed with apologies]

These passages long ago attained their well-deserved immortality in the record of human artistry. Not only are they perhaps the best description in all of literature of the poignant permanence of things lost to us forever—hence of that gift of memory upon which hinges the very life of the soul—the passages are themselves exquisite universal "madeleines," evoking once more into life for every reader his own recollection of what it is like to forget matters long dead and then to have some innocuous small thing awaken them once more into vivid existence.

There is almost no higher brain activity that does not depend on memory; and many activities that we conceive loosely as some separate faculty are in truth merely variations of remembering. Pattern recognition, for example, the chief activity in which artificial neural networks are now being applied to supplement human intuition—or supplant it—is such a variation: To place an object in a category is first to memorize the categories, then to recall the correct category so as to match an object to it.

À LA RECHERCHE DU TEMPS PERDU
—FROM MADALINE TO MADELEINE

But Proust's wonderful portrait of recollection as we experience it reveals an important nuance that transcends mere categorization and points toward the true nature of human memory: When we recall something, we do not merely identify correctly the categorical type of which it is an instance. Rather, we hold in our mind's eye some part of a larger experience—the madeleine—and this mere part, evidently having in itself sufficient critical detail, evokes the whole of the experience. For a time, the whole of which the "trigger" is part may indeed remain obscure, hidden, not yet assembled, in some sense not yet even created, as Proust observes. But slowly, surely, the mind converges on the recollection; bit by bit, at first with excruciating uncertainty and slowness, the memory builds; each new element, as it settles into its proper place, then serves to evoke yet other parts of the whole, faster and faster, until in a final crescendo that happens too quickly to time, the "vast structure of recollection" tumbles at once into sight: whole, intact, untouched, alive as ever, however long until that moment forgotten.

This kind of memory, so unlike the numbered storage bins of standard computation, is surely beyond the capacity of any neural network, it was thought.

In 1982 neural networks were still in deep eclipse. While pioneers such as Teuvo Kohonen and a few others in the United States continued to pursue them, it was tough going everywhere. No matter how outstanding their contributions would later be seen to be, any scientist spending his time on neural network research was thought to be foolishly leaving the herd. But herds, stampeding irresistibly in one direction, will suddenly reverse course. So it was with neural networks.

JOHN HOPFIELD AND THE HOPFIELD NET

In 1982 a Cal Tech particle physicist with an outstanding academic reputation, a patrician personality, and an especially elegant way of presenting complex ideas delivered a lecture on neural networks at the prestigious annual meeting of the National Academy of Sciences. That paper was, in fact, a masterstroke of physics. His talk on so outré a subject electrified the audience,[1] a large number of whom were neither computer nor brain scientists but physicists. John Hopfield demonstrated that a certain kind of neural network is mathematically identical to a certain kind of magnetic system, a so-called spin glass. Spin glasses were not only well accepted, they were an object of intense fascination for physicists (if for no one else). Neural nets that mimicked these magnetic systems, oddly enough, turned out to be not only simple to understand, but they formed maps and stored memories just as did Kohonen's nets, and they were biologically realistic in requiring no outside supervision to work: They, too, organized themselves. That they mimicked a system found in nature—a simple, inorganic system, to boot—suddenly made it seem completely reasonable that such methods might have been adopted by nature in shaping the brain—no outside supervisor, after all, had created natural magnetic ores. Hopfield had done an end run around the bogeymen that had been spooking the herd for almost twenty years and turned its rebels and reluctant stragglers into its leaders. "The theoretical physics community is like a swarm of locusts," as one of the new leaders put it. After Hopfield's presentation of neural networks, "all the physicists landed on it."[2]

IRON SCOUTS

Here's what Hopfield's idea was about. Spin glasses are magnetic substances that demonstrate so-called collective behavior. Collective behavior refers to the fact that an aggregate of many identical elements can sometimes generate large-scale coordinated patterns in the whole collection, even though there is no outside "orchestrator." Mutual effects between adjacent elements (only "local interactions") generate the global patterning.

A band, for example, does not display "collective behavior" in this sense, since the behavior of all the individual instrumentalists is in large part guided by a conductor. A colony of bees, on the other hand, does demonstrate remarkably intelligent collective behavior, since the order it displays arises almost entirely out of the local interaction of small numbers of individual bees. (The queen plays an important role, but not as micromanager of the colony's activities.)

Now consider two rectangular magnets side by side, both free to rotate, their north poles both pointing in the same direction, magnetic fields influencing each other. (See figure 6–1a.) This mutual influence is somewhat like two neurons with reciprocal connections onto each other.

If we jiggle the magnets, what will happen? Even though both are free to rotate, after a bit of vibrating, the magnets will settle back in the original configuration, parallel to each other. The two north poles repel each other, somewhat like inhibitory connections, but in the case of magnets, which always

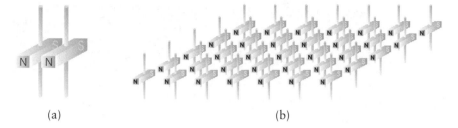

FIGURE 6-1   (a) Two magnets cheek by jowl. (b) An array of magnets.

have both north and south poles, so do the south poles. The ends repel, but if the two north poles move apart, the two south ones move together and repel even more strongly. What happens if we set up an entire array of such magnets, as in figure 6–1b?

Once again, the stable "equilibrium" position for such an array of magnets is for them all to line up in parallel. . . . Well, not quite. Closer to the edge of the array a magnet is not subject to perfectly balanced forces. If we look at the array top down (see figure 6–2a), we see that the magnets near the edges of the array are subject to unbalanced forces.

Now, what happens if we completely flip a couple of magnets into the exact opposite position (maybe by jiggling extra hard)? Surprisingly, they won't return to the original position (unless we jiggle extra hard again, and they happen to, but, then, maybe others will flip). Instead, they will introduce a distortion into the balance that is transmitted from one magnet to the next. In the end, the array will once again settle into a final, stable position toward which, after any minor jiggling, it will return. But it is not the exact same arrangement as before, since many more magnets are now aligned not in parallel, though still in orderly fashion, as shown in Figure 6–2b.

Any "ferromagnetic" material (e.g., iron) is composed of atoms, the direction of whose electrons' spins act very like an array (a three-dimensional array) of tiny magnets. Neighboring atoms in a ferromagnetic material tend to cause

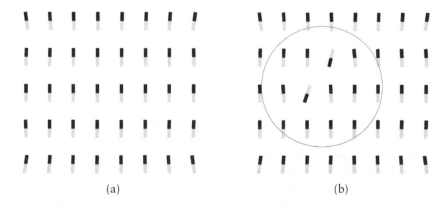

FIGURE 6-2   (a) An array of magnets, imperfectly balanced at the edges. (b) A new arrangement caused by two flipped magnets.

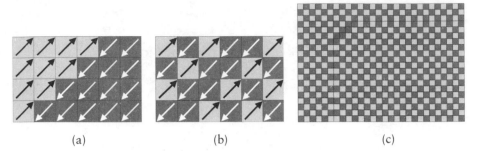

FIGURE 6–3   (a) Electron spins in two domains of common alignment. (b) "Anti-iron" spins in two domains of reverse alternation. The boundary consists of a double diagonal band of down-and-to-the-left spins (white arrows). (c) More "anti-iron" spins in two antiferromagnetic domains. The diagonal segment of the boundary in this illustration is the one shown in (b). Note that the balance of forces differs depending on its direction.

each other's electrons to align together, but if a region gets established with the reverse alignment, it, too, tends to persist. The result is many tiny domains of various electron spin orientation, hence regions of different magnetization. (See figure 6–3a.) A large piece of iron is therefore not magnetized as a whole—regions of competing magnetizations cancel out.

The domain boundaries, where spins held in one orientation sit close by spins held in the opposite direction and are able to influence each other, are of especial importance. The balance there is tense. A slight change in external forces and spins can flip, causing subsequent ones to flip, like agents of the CIA and the KGB socializing cheek by jowl at Embassy Row parties in the Gorbachev era. A wave of spin flips can propagate, causing the domain boundary to travel. Later we will see how this phenomenon relates to the propagation of signals. But of greater importance for relating neural networks to spin glasses is the idea that in an array such as this, certain elements can be poised at the verge of changing.

ANTI-IRON

There is another kind of magnetic material, less well known to us in our day-to-day experience than common magnets. In these so-called antiferromagnetic materials, spins tend to align in *alternating* directions. Now, above, we made a quick and, as it happens, overly simplistic analogy between attractive magnetic forces and excitatory neuronal connections on the one hand and between repulsive magnetic forces and inhibitory neuronal connections on the other. The subtler, more precise analogy is that a ferromagnetic group of neighboring spins that tend to cause each other to align are mathematically exactly like excitatory connections between neighboring neurons; and a group of neighboring spins that cause each other to anti-align—to flip rather than reinforcing their current status—are like inhibitory neurons. Left to itself, antiferromagnetic materials develop spin domains where the order of alternation itself varies from region to region.

In this example, too, there are two domains of (this time alternating) spins, though the boundary is hard to see. With more spins represented (see figure 6–3c), the boundary becomes evident.

We now need to take these spin analogies to one more level of complexity—by mixing ferromagnetic and antiferromagnetic domains to create a material called a spin glass. This analogy is so powerful for the following reason: An array of ferromagnets subject to a magnetic influence from the outside can store a memory—that's exactly what happens when we use, say, Memorex audiotape to record a song. The tape is brown because the recording surface is made of iron. But this kind of tape (like tape made of some antiferromagnetic material) can store only one memory at a time. In storing a new memory you erase the old one. A spin glass, however, can store many memories all at once, just like a neural network and like the brain, and it does so spontaneously, without being given any instructions. It is able to do this because a spin glass array of magnets and antimagnets has more than one "best" arrangement. Each such best arrangement can get connected to a different memory and store it. Here's how.

FRUSTRATION, THE MOTHER OF INTELLIGENCE

Spin glasses are composed of a random mixture of both ferromagnetic and antiferromagnetic material—a random mixture, as it were, of inhibitory and excitatory connections between neurons. Adjacent spins in a spin glass can compete to either align or flip their neighbors. The balance is so close that the materials are referred to as *frustrated*—the spins are always on edge, ready to flip at the tiniest jostle. (Since the silica molecules of cut crystal are all neatly aligned, while in everyday glass they're helter-skelter, a frustrated system of spins is called a spin *glass*.) One neighborhood of four subatomic magnets subject to both kinds of influences interacts as do four neurons with a specific arrangement of reciprocal connections. (See figure 6–4a and b.)

In figure 6–4a, spins 1 and 2 are both up, and they are "coupled" to each other "ferromagnetically." That is, each influences the other to align in the same direction as itself. The directions of both are therefore consistent and mutually reinforcing, whichever it happens to be—up or down. This coupling is equivalent to the reciprocal effects of the two excited neurons (white) shown in figure 6–4b, with mutually excitatory connections.

Now, we also see that spin 2 is up while spin 3 is down. They are aligned oppositely, as it happens—their spins alternate. Furthermore, the coupling between them also happens to be "antiferromagnetic." As a result spins 2 and 3 each tend to cause the other to align oppositely to itself. Both their states as shown are therefore likewise consistent and mutually reinforcing. Hence, spins 1, 2, and 3 form a consistent and mutually reinforcing triad. The same applies to the corresponding neurons in figure 6–4b.

So, based on this configuration, in what spin state (state of excitation) is spin (neuron) number 4? Spin 4 is frustrated, unable to "decide" which way to go, for there exists no state that satisfies all the constraints placed upon it by its

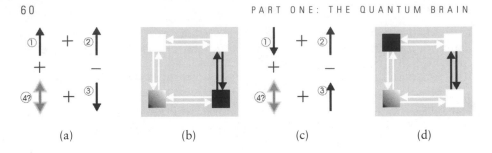

(a)   (b)   (c)   (d)

FIGURE 6–4   (a) A neighborhood of four spins compared to (b) four interacting neurons forming a small network. The two possible spin directions (up or down) are represented by the directions of the arrows. The two possible neuronal states (excited or quiescent) are represented by the colors of the squares. The two possible kinds of interactions between spins (ferromagnetic or antiferromagnetic) are represented by the + or – signs between them. The two possible kinds of interactions between neurons (excitatory or inhibitory) are represented by the colors of the arrows connecting them. (c) A different neighborhood of four spins compared to (d) four interacting neurons.

neighbors. Spin 1 is up and couples to 4 in such a way as to make 4 up, as well; spin 2 is down and couples to 4 in such a way as to make 4 down, as well.

Note, too, that if spin 4 were to decide to stay up, then spin 3 would be frustrated; if spin 4 were to decide to stay down, then spin 1 would be frustrated. If spin 4 and 3 decide to stay up, then spin 2 is frustrated; and so on. So long as three out of four couplings are ferromagnetic and one is antiferromagnetic, no arrangement of spins is perfectly stable. Instead, there will be never-ending competition among equally imperfect states.

However, the solution shown above and the other equally bad solutions are not the worst possible. Indeed, they're the best possible, hence equally good. For example, let's say that the four spins (with the same set of couplings) were arranged as in figure 6–4c & d.

We see that spins 1 and 2 are in inconsistent states, as are spins 2 and 3, with 4 again frustrated. Left to itself, this network of spins (or a network of four neurons with connections as shown) may not be able to settle into a perfectly stable state, but it will settle partially into one (or rather a never-ending succession of all) of the least-bad (= best possible) states.

The process of "settling" or "converging" in a such a network of spins (or neurons) is akin to the trajectory of a ball released from some point high up on a bumpy landscape. In fact, it is mathematically identical to it. We know that of all the places the ball might settle, it won't settle on a peak or on a slope, but in a valley. If there are many attached valleys all at the same level, none of these will be favored, and (assuming there's no friction) the ball will drift aimlessly from one spot to the other, as in figure 6–5a.

If there were a single solution—a single most energetically favorable configuration: the most settled arrangement—the network of spins or neurons, or a rolling ball, would seek it. And so long as this configuration was not surrounded by impassable energy peaks, eventually it would be found. The simplest example would be a pit, as in Figure 6–5b.

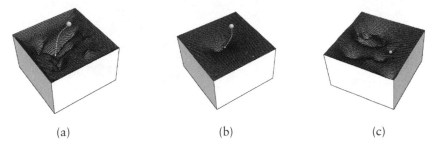

(a) (b) (c)

FIGURE 6-5 (a) A system seeks the most favorable—that is, the lowest accessible—energy state. More than one position may be equally favorable and equally accessible, as with a frustrated network of spins or neurons. This frustration may be represented as a branching network of valleys all at the same (lowest) elevation. There exists no perfectly stable location—a pit—into which the ball can fall and stop moving. (b) In this case, there is only one lowest energy state, and it's readily accessible. That's the configuration in which the system comes to rest. (c) Energy landscape with many "basins of attraction." The ball is shown at rest in one, not the lowest. This is a "local energy minimum," not the "global energy minimum."

For obvious reasons, these energetically most favorable states are referred to as "basins of attraction."

In the above example, there is not only one lowest point on the landscape, there is only one basin. In actual spin glasses, of course, there are many, many more such spin networks, all linked together. As a result, there are a great many different such basins of attraction, at various energy levels, and most are not linked together. Hence depending on the precise sequence of spin flips (the precise path taken by the ball), the network may settle in one or another basin. The basin it settles into may not necessarily be the absolute lowest—note the "volcano crater" in figure 6-5c—just low enough so that to leave it, it would need to go from a more favored to a less favored state. This it won't do, unless it is "jogged." (We'll see shortly what "jogging" means in this context.)

Now, what do these basins of attraction have to do with memory? The principle may be stated simply: *Each basin in a frustrated spin system corresponds to a memory.*

We are presented with a stimulus. It may be a fragment of the whole picture, or it may be a distorted version of the whole or of a part—*No sooner had the warm liquid mixed with the crumbs touched my palate than a shudder ran through me and I stopped.* This corresponds to the ball being placed at a particular spot on the landscape, relatively close to the bottom of a basin, or not, as the case may be. Some external force has impinged upon the network (of spins or of neurons), stimulating it. It starts to reconfigure.

Spontaneously the input is processed, shuttled back and forth, altered. These are the delicate, tense moments before we recall, but when we know the memory is there—*I feel something start within me, something that leaves its resting-place and attempts to rise, something that has been embedded like an anchor at a great depth; I do not know yet what it is, but I can feel it mounting*

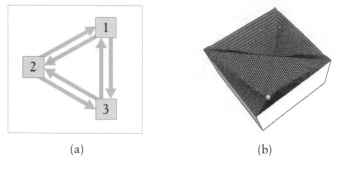

(a)   (b)

FIGURE 6–6   (a) Canonical microcircuit as a three-neuron (spin) network. (b) Energy landscape for the simplest (three-spin) Hopfield network.

slowly; I can measure the resistance, I can hear the echo of great spaces traversed. These moments correspond to the ball moving with increasing speed, spiraling toward a basin, the network of spins, each "on edge," reconfiguring itself, ever closer to a stable state; the neurons of the neural network passing excitation and inhibition back and forth, searching for a stable configuration.

Finally, the input pattern is converted to one of the stored patterns, the memory "clicks," the ball drops into the lowest energy spot of the basin, the initial stimulus is transformed into a remembrance of something past—*And suddenly the memory revealed itself. The taste was that of the little piece of madeleine.* The network's output is its final configuration, the relatively stable new configuration it adopts, until another stimulus prompts into another reconfiguration.

If the initial stimulus is a fragment of the complete memory, then the final configuration will be the complete memory and so will bring in its train many additional elements not recalled at first. *Immediately the old grey house upon the street, where her room was, rose up like a stage; and with the house the town, the Square, the country roads we took.*

Human memories are retrieved by *similarity*, not by address, since in a network there are none: Every image is distributed everywhere. This *associative recall* is demonstrated by networks (whether of neurons or of spin) of the utmost simplicity. Even three elements, the basic building blocks of the fourth brain, the canonical microcircuits discussed before (the equivalent of a three-element spin glass), can memorize, a variant of which is shown in figure 6–6a. In fact, *it is the smallest such unit that can*—specifically, it can store and subsequently recall two fundamental memories. (Appendix A explains the details of how this works.) Larger, more complex networks can then easily be built up out of many such basic units.

Using the energy diagrams of figure 6–6a, a miniature spin network/microcircuit has two basins of attraction each surrounded by three discrete positions—see figure 6–6b.

Before such a network has been trained (before it has stored any memories, or patterns), the landscape would be almost flat. As two objects are presented to it—apples and bananas, for example—it quickly forms two depressions and associates one with each object. Each depression represents one of the two lowest-energy stable states into which the network will settle, depending on where it starts—that is, depending on which pattern of stimuli is fed in from the outside. It's easy to see that a large network composed of a great many of these subunits would function very much like a network of binary (i.e., on/off or 0/1, or −1/+1) elements. In particular, two subunits passing opposite states back and forth would form a two-state oscillator; additional units will conceivably establish multistate oscillators.

To be useful, memories of this sort—associative memory—must have many more than three neurons (spins), and their dynamics and energy landscapes quickly grow exceedingly complex. Yet their performance remains much the same in principle. However, with only three neurons, a network will memorize its two items perfectly. As the number of neurons increases, a Hopfield net will develop the capacity to make mistakes—to fuse parts of different memories and so create unreal ones—false memories, or even "hallucinations." Think of these as extra valleys, if you will, formed by the intersection of two others. If the network ends up in one of these "spurious attractors," as such unreal basins are called, its output will be something it has invented. Even without such spurious attractors, there is always a chance that the ball will roll downhill with enough speed to seek out a low, but not the lowest, spot in the proper region—the memory will be only partially accurate. To develop neural networks that make few errors in solving complex problems, and to better understand how the brain does likewise, something more is needed.

TURNING UP THE HEAT

Suppose a network—whether of spins or of neurons—does get stuck at a spurious memory, like Maurice Chevalier recalling dinner at nine. What happens when such a memory gets "jogged" by a more accurate Gigi—"We met at eight"—and suddenly falls into the correct place? ("Ah yes! I remember it well.") If we think of a ball searching for the deepest nearby basin of attraction, then a "jog" corresponds to kicking the side of the box to jiggle it a bit— just enough to jar it out of a spurious basin and continue its downhill search for a fundamental valley. To increase the likelihood that our ball will find its way into a fundamental basin and avoid shallow ones, we would try to keep it constantly jiggling as it rolled—say, by placing the landscape on a tabletop vibrating *just so*: Too little vibration, and the ball could remain trapped in some local rather than *the* global minimum; too much and it will jump *out* of the genuine one, never settling anywhere.

In living networks, the equivalent of a vibrating tabletop is simply the random jiggling in place of atoms and molecules—that is, *heat*. The energy that maintains our body temperature at 98.6° is used by the brain to keep neurons in a constant state of readiness to fire. By adding artificial heat to an artificial network, you introduce a degree of uncertainty into the "motion" of the network as it searches for a minimum. This you do by making the output of each neuron partially random, or *probabilistic*. Such neurons are called "stochastic," meaning "appearing random."

For example, say all the inputs add up to 10, and the neuron contains a rule (multiple thresholds determining the output) that does not merely say "If 10 in then 20 out." Instead, you roll some dice and have the neuron put out a signal that falls between, say, 18 and 22, with 20 being the most likely output, 22 the least—a "bell-shaped curve" of likelihood centered on 20. Such a procedure not only helps Hopfield-type networks not get stuck in spurious attractors, it enables them to more likely reach the true (absolute) energy minimum. The fact that our own living neurons are also stochastic helps us better avoid distorted (or altogether false) memories and in general to recall better and faster. Once again, nature has taken advantage of the "messiness" of a natural environment to enhance intelligence.

By affecting its neurons, "temperature" also affects the overall performance of a Hopfield network. Recall that in human memory, experiences are not stored according to some kind of address-based filing system ("kindergarten memories are located beginning at address CA47-008796-AY and continuing until CA48-744438-FD; last defragmentation 6/1/99; FAT table updated 6/1/99 . . .") but by association—similarity of form, contiguity in time, proximity in space, emotional category, and so on. Sometimes associations are mere auditory puns based purely on sound: thinking of "caboose" may evoke the image of a moose. Others are strongly colored by emotion: thinking of "Oswald" may evoke the entire history of the Kennedy era.

Early psychologists discovered that associations form clusters, and they developed the famous "word-association test" to probe the way clusters were structured. When a stimulus comes in (e.g., a word), it may be thought of as located more or less "near" the memory to which it is going to "travel." Those experiences that are both universal and powerful will develop strongly reinforced "valleys" (fundamental memories) that tend to be found in everyone—all the more so if the experiences are built up around primary human instincts and needs (third-brain objectives).

For instance, the image of a large-breasted naked woman for some people subtly evokes the long-forgotten emotions associated with mother, just as for Marcel, the narrator of *Swann's Way*, the combination of tea and madeleines subtly evokes long-forgotten emotions associated with his aunt. Points on the surface of an energy landscape that feed the ball into a given basin of attraction represent experiences that are more or less closely associated with that memory. (The experience may be past or present, real or partially imagined. In any event its representation in the brain is as a configuration into which the network is provoked as a result of the incoming stimuli.) Consider one of the landscapes we showed before, both rotated forward in three dimensions, in figure 6–7a and as a two-dimensional topological map in figure 6–7b.

We see that different positions on the map more or less strongly belong to one basin or another. Since the shading is done in steps, boundary regions can belong to more than one basin. At a finer level of detail, boundaries are sharply defined—*if* the neurons are absolutely deterministic. If they're probabilistic, however, boundaries are truly fuzzy: Given two identical inputs, the output can vary.

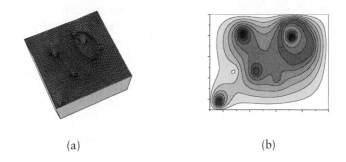

(a)   (b)

FIGURE 6-7   A typical energy landscape. (a) Figure 6–5c rotated forward so that we may more easily compare it to (b), the same landscape shown as a topological map coded by height.

"Heat" continually pumps energy into the network, giving its trajectory an "activated" quality—like a pinball hitting driven bumpers or being flipped by the player. The network can therefore leave lower energy states to access higher ones. Compare the deterministic trajectory in figure 6–8a (left) to the probabilistic one in figure 6–8b (right).

On the left in figure 6–8a, we see that the network might have settled in B but is diverted by a small high-energy peak and curls around A, skirting the attractive domains of C and D. It settles quickly at A. In human terms:

"Who does she remind you of?"

"Umm, my mother!"

"Anyone else?"

"Well, she looks vaguely like my cousin, Christine, I suppose, or maybe Mar . . . no. No, not really. My mother."

E is inaccessible because, even though it is itself a stable, low-energy spot, it's surrounded by a closed "wall" and the walls by a closed "moat." Keep in mind that by "low energy," we don't mean anything like "lethargic." We mean

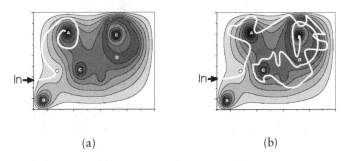

(a)   (b)

FIGURE 6-8   Trajectories in an energy landscape as a system settles into a basin of attraction. (a) Deterministic network. The input stimulus ("In") evokes and becomes associated with memory A only. (b) Probabilistic network. The input stimulus ("In") evokes and becomes associated in varying degree with memories A, C, D, and E.

a memory that is favored, grooved deeply, as it were, and easy to reach—usually. A memory that is deeply embedded and powerful but "blocked," as this one is, will not be easy to reach, as if it were barricaded. When such barricaded memories are reached, often after an investment of much time and energy to scale the walls, we have a subjective feeling of "How did I ever forget that?" Here, because of the repelling walls and attractive moat:

"Don't you see the resemblance to Greta Garbo?"

"Garbo? I can't see it at all."

If we add heat (in this case, a lot of heat) to the network—in a human being, this can be done via anything that "loosens" associations and speeds up the flow of ideas (e.g., chemicals, instinctive excitation)—the results are very different:

"Who does she remind you of?"

"Well, she really looks a lot like my mother, but you if you sort of squint you can see the resemblance to Greta Garbo, who I once saw on a train in Klosters, or was it Davos?—Garbo. Right . . . nah . . . no—well, yeah, but n— in any case, I remember that movie—what was it? Oh, yeah, *Queen Christina*, and you know, now that I think about it, she sort of looks like my cousin Christine who also went skiing in Klosters once where that friend of Garbo lived, I think, in any case, yep, she's the spitting image of Garbo, who I think is absolutely fantastic, I mean, like once I . . ."

"Not your mother?"

"My mother!? What? C'mon, you're nuts."

In the first case, the mind processes some initial stimulus in a straightforward way, moving quickly toward some essential memory that the stimulus "reminds" it of. In the second, more energetic case, the mind senses similarities among the stimulus and many different memories. It creates a wider web of associations. The overall "energy level" of the net—the relative flatness or incline of the probabilistic response—determines to what degree it makes narrow associations or wide ones. Clever people—and maniacs—are prone to the latter.

THE LUNATIC, THE LOVER, AND THE POET ARE OF IMAGINATION COMPACT

In his famous essay on the art of poetry, Aristotle makes a keen observation: "But the greatest thing by far is to be a master of metaphor. It is the one thing that cannot be learnt from others; and it is also a sign of genius, since a good metaphor implies an intuitive perception of the similarity between dissimilars."[3]

To first see similarity where others see none: This is precisely what happens in the trajectories of highly energetic networks. Widely disparate "locations" (estates) are linked together in a single train of thought. Alternatively put: More states are accessible in the quest for the best. The capacity for exploration—for entertaining linkages that might prove to be wild—is the heart of mental play. The genius toys for hours on end, exploring the craziest possibilities. Only after playing does he yield to the rational exercise of top-down logic and to judgment based on evidence. Only then does he reject wild hypotheses that prove false; refine and recompose wild hypotheses that prove true. As Feynman said of the mathematician Srinivasa Ramanujan and of himself, when pressed to explain their genius: "Not so much the ability, but the desire to play around . . . I've always played around. . . . I was just playing, like a child playing, but with different toys."[4]

So, if you want an artificial network to grow "smarter," you might consider squashing its input/output curves; make it play around more before "settling." But does the same hold true for a natural neural network? Can a human brain be made artificially smarter by affecting its neurons?

Paul Erdös was one of the century's deepest and most prolific mathematicians. In a field where highly accomplished practitioners may publish no more than fifty papers in a lifetime, Erdös published almost fifteen hundred, "many of them monumental and all of them substantial." He was also a habitual user of brain-altering chemicals: ". . . for the last twenty five years of his life [Erdös died at age eighty-three] . . . he put in nineteen-hour days, keeping himself fortified with 10 to 20 milligrams of Benzedrine or Ritalin, strong espresso and caffeine tablets. 'A mathematician,' Erdös was fond of saying, 'is a machine for turning coffee into theorems.'"[5]

Is that all a genius is, then? A kind of semicontrolled speed freak of nature, who may just as well up the chemical ante himself? There's another side to the story. Push it too far, and you end up with a network that is not so much brilliant as bonkers. Among Romantics, this is known as the "divine madness of genius"; among computational neuroscientists, the "stability-plasticity dilemma."

We should be properly awestruck when a Shakespeare lights upon "Juliet is the sun," forever impressing on the imaginations of men the unutterable, life-giving illumination of first love in a way that no other combination of words could better convey. But we are not so much impressed as saddened when the bedraggled old lady, now going on forty years in a mental institution, announces about her own lost adolescent love from Italy, "Naples and I must provide the whole world with noodles."[6] What's the difference?

We see at once how Juliet is indeed the sun. In that moment, our understanding of love is altered. (Our understanding of the word "is" is altered, as well—though not in the way become recently notorious.) We grasp this while at the same time understanding perfectly well that a thirteen-year-old girl is not, in fact, a gigantic, 93-million-mile-distant fusion reactor. However far apart the two terms of Shakespeare's metaphor—young girl, nearby star—he has brought them together in a way that is self-evidently true, that therefore requires no further explanation. The good metaphor is diamantine: The absolute minimum-total-energy trajectory linking two separated points that lesser minds cast about for and never find—or find instead its zirconium variant: because they foolishly conclude that "All is one."

Yet in the course of arriving at the minimum, the great mind in its play produces absurdities, as well—by the cartful. We are not privy to Shakespeare's scratchpad, let alone his naked thoughts; we have no idea how many cloying similes equal to or worse than our madwoman's were crossed out and discarded, or never committed to ink, before he settled upon "Juliet is the sun."

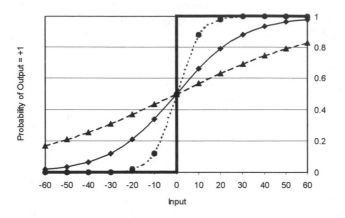

FIGURE 6-9   Genius or madness? Different probability functions for neurons in the fourth brain:

• Relatively steep function corresponds to normal, "uninspired" state.

■ Shallower function corresponds to "inspired" state of "intuitive perception of the similarity among dissimilars": Genius.

▲ Very shallow function corresponds to excessively energetic network: wild associations typical of "mania." Treatment with lithium or anticonvulsants steepens the curve once again.

Shown in bold is the fixed, steplike input-output function of the original Hopfield-type network.

Perhaps he first thought to have Romeo declaim "Juliet is my main dish," because at that moment, idiosyncratically, he was himself both lovelorn and hungry.

But in the madwoman's makeup, no ability to separate sense from nonsense remains. Her permanently overheated network can no longer cool.

When the energy of an artificial network gets too high, disorder ensues. Random links are made, they dissolve and give way to others, equally wild, they reform anew without stop—just as happens to those who push their use of "speed" beyond a certain point, or to manics.

The solution to this problem in artificial networks is relatively easy—turn down the "heat": Steepen the probability function to make it less stochastic, more rigidly deterministic. Make the firing more predictable, less random; remove or diminish the "hair trigger." Is there an equivalent for the natural networks of the fourth brain?

Lithium is the most well-known antimanic agent, and its effects on neurons are still somewhat mysterious. But the other class of antimanic agents happens to include the same medications used to treat seizures. That is, they are drugs that tend to reduce the firing susceptibility of neurons and the tendency of a cluster of neurons to fire wildly in response to the firing of others. Different such susceptibilities are compared in figure 6–9.

An unusually large number of individuals who suffer from mania are in fact also extraordinarily creative, but there's a catch: As the associations grow more and more distant, the very judgment to discard excess becomes impaired. Everything is now perceived as exalted, nothing as execrable. The state of inspiration feels wonderful, so the more inspired one feels, the more reluctant one becomes to give it up—and the less able to judge that it ought to be given up. Mania tends to expand, making it more likely that it will expand yet further.

Lithium, on the other hand, brings with it improved judgment, but at the cost of feeling less inspired. Creative individuals who are prone to mania often waste many years in an effort to game their medications, aiming for inspiration without impairment. But it's a devil's bargain they seek: The minimally acceptable level of inspiration is almost invariably coupled to impaired judgment. Acceptance of what feels like a limitation on one's creative powers usually therefore comes only after a long and agonizing struggle and the repetitive payment of severe damages.

Kay Redfield Jamison, a brilliant professor of psychiatry at Johns Hopkins, a gifted writer, as well, and herself a sufferer from manic-depressive illness, describes the dilemma both she and her manic patients find themselves in:

> My manias, at least in their early and mild forms, were absolutely intoxicating states that gave rise to great personal pleasure, an incomparable flow of thoughts, and a ceaseless energy that allowed the translation of new ideas into papers and projects. Medications . . . cut into these fast-flowing, high-flying times. . . .
>
> [Later], I was manic beyond recognition, and just beginning a long costly personal war against a medication that I would, in a few years' time, be strongly encouraging others to take.[7]

The "stability-plasticity dilemma" or the "divine madness of genius" turns out all too often into the battle over "lithium noncompliance."

TWO ROADS PARTED . . .

Associative memories such as the Hopfield network are ideally suited for solving problems that go well beyond memorization and recall. These are the so-called optimization problems, where a wide variety of similar solutions to a task are available, but there is a high premium on finding the best possible one in a limited time. The classic example is the "traveling salesman," as shown in figure 6–10.

These kinds of problems easily become so complex, in fact, that for all but the simplest it is not mathematically possible to prove that one has arrived at the best solution. A guarantee that one is near the desired "extremum," within a certain margin of error, is the best one usually gets, and often not even that.

Evolution itself can be treated as a enormous, multibillion-year-old optimization problem, "minimum energy" being recast as "optimal adaptational fitness." The Hopfield net is much like nature, since, so far as we can tell, all of

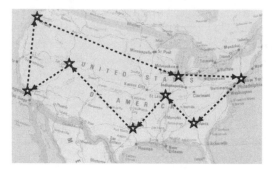

FIGURE 6-10   Traveling Salesman problem. Arkady arrives in New York and must depart from there. He must visit Chicago, St. Louis, Atlanta, Dallas, Salt Lake City, Los Angeles, and Portland. What route is fastest? Cheapest? Safest?

nature operates according to energic principles. Indeed, all of nature—not just the biological section—may be considered a gigantic optimization problem since all of nature seeks out the path of "least action," a specific form of energy minimum. Nature not only seeks energy minima, but seeks them along paths that minimize the sum of the difference, at every point along the trajectory of a system, between kinetic and potential energies. When this principle was first discovered, in the eighteenth century, it was thought by many to provide evidence for a divine intelligence behind nature.

Nature seems to have devised an extraordinary bag of tricks to solve its various optimization problems. Given what we know so far about the brain, and given the many optimization problems it has always been extraordinarily clever at solving (long before mankind had a name for what it was doing), it seems clear that as a computer, the brain is an optimization device—the subtlest, deepest, most complex such optimization system in the known universe.

Given the limitations evident in the Hopfield network—its need for almost universal connectedness, its inability to attain with certainty the extremum it seeks—we may be certain that nature has quite a few additional tricks up her sleeve. Heat alone is insufficient.

A TIME TO ANNEAL

Bend a paper clip back and forth. The metal gets both stiffer and more brittle. Eventually it snaps. A paper clip is relatively soft and malleable ("ductile"). But as the metal is "worked," it grows harder and starts resisting, until it gives, releasing a burst of energy all at once. The smithy's art depends on this phenomenon. To make a sword's edge, metal must be hammered and bent repeatedly, so it must be soft and malleable. But as it is worked, it hardens and resists further shaping. The smithy thus needs repeatedly to turn brittle metal soft and pliable again, while still retaining the shape into which it has so far been worked. If the smithy succeeds, he can work it further.

The smithy's secret lies in applying both heat and cold; in when and how much heat he applies to the steel, and when and how he allows the metal to cool. The process, called "annealing," works roughly as follows. The metal is malleable to begin with. After the swordsmith hammers and shapes it for a while, it gets too hard and brittle to continue. So he heats the sword just so: enough to soften it, but not so much as to melt it. In its now once again malleable state, the smithy shapes it further. As he hammers, the metal cools and from the working slowly becomes brittle again.

A sword needs a fine edge. As the hammering flattens the edge ever thinner, smaller and smaller amounts of heat will cause the edge to melt entirely. So the smithy uses a finely tuned "schedule" of alternating heating and cooling. As he gets closer and closer to the final product, the amount of heat (or cold) he applies gets smaller and the cycles of hot and cold get shorter.

The last act of the swordmaker—immortalized in innumerable Hollywood films—is suddenly to quench the heated sword in cold water to "temper" it. What happens here is roughly as follows. Metal is composed of innumerable, tiny ricelike grains of crystal. When a metal is soft and ductile, the grains are all aligned. When hammered, domains of different alignment are created, causing fracture planes. This makes the metal hard but brittle—prone to snapping. A sword must be hard but not brittle. This happens when its grains are randomly aligned—no continuous boundaries between domains, yet the grains are unaligned. This state can be achieved by heating the completed sword just so and by quickly cooling it—quenching it—just so: very difficult to do well.

A malleable piece of metal is in a very low energy state. A master swordsmith has learned the art of putting metal into an extremely *high* energy state, where the energy is distributed with great uniformity; he has balanced a ball on the tip of a narrow peak. The sword's edge can be made exquisitely thin without chipping; it resists bending without snapping; if it does snap, you don't want to be nearby.

If the goal is not a sword but a piece of malleable metal like a paper clip, quenching is avoided. Instead an *annealing schedule* (hot-cold-hot-cold . . .) is sought that will maximize the "search" of the metal atoms for the best possible state of grain alignment. We can improve the performance of neural networks—whether naturally occurring or artificial—in their search for an energy minimum by similar means.

If we merely introduce heat into a network (we make the neurons respond to their input with a somewhat random output), we pay a price. Namely, while the jiggling of the ball will make it more likely that the ball won't get trapped in some relatively high-energy minimum (a volcano crater high in the mountains), it also makes it more likely that the ball will fail to come to rest in an absolute minimum (Death Valley).

To improve the intelligence and reduce the error rate in artificial neural networks of the Hopfield type, we *simulate annealing*. We start the network off in a relatively "hot" state—its neurons respond with a lot of randomness to their input—then cool it in steps. At higher temperatures, the network is not likely to settle in or near an absolute (global) energy minimum, but it is also less likely to get stuck in or near a high-energy relative (local) minimum. If the temperature is now turned down a bit, some or all of these high-energy local minima will become inaccessible. (In the energy landscape picture, the jiggling ball no longer bounces around enough to climb back up, against the force of gravity, into one of these undesirable states.)

But if the ball is still energetic enough to keep going without getting stuck, it is likely to avoid the local high-energy minima that are present at its new, lower level.* It may be shown mathematically that, for any given type of energy landscape, there exist annealing schedules which will ensure that the final state of the network is either at or very near the global minimum. (That this must be the case we know from the fact that metals can be annealed.)

Simulated annealing in Hopfield-type networks with "stochastic" neurons has become the most widely used way to apply these networks to real-life problem solving. They form a very close analogy to how human networks work best, too. In a well-structured creative process, whether individual or group, the best initial state is one of relatively high temperature: lots of opportunity to generate ideas, crazy or otherwise, without constraint; one notion stimulating others without critique of any sort—brainstorming—a state that favors the creation of new ideas (the fresh perception of similarity among dissimilars), both good and bad.

There then follows a period of cooling, when the initial ideas are consolidated and pruned, but with a light hand. Individual ideas that remain are expanded on in minibrainstorm sessions. This is then followed by even more cooling, and so on. When an individual brainstorming in a group can't cool down at the right time, the group will cause her to chill.

Hopfield networks and their progeny thus provide a compelling model for how the fourth brain works, mistakes and all. But they, too, have problems, both with respect to how much they can do well and with respect to their biological plausibility. For example, they require an enormous number of connections. Indeed, their basic principle is "every neuron connected to every other." However densely interconnected is the fourth brain—amazingly so, as we've seen—it is not completely interconnected. Imagine what that would mean: 20 billion neurons each with 20 billion connections. Even divided into subunits as it is, universal connectedness is simply not the case.

Between 1982, when Hopfield's presentation turned computer and neuroscience research on its head, and 1998, a race ensued to develop ever more sophisticated variants of the Hopfield net, and ever more biologically realistic ones—ones that don't require such massive interconnectedness. This last challenge has proven the most difficult and only very recently have the final pieces been put into place: It turns out that if a large number of *oscillators* of a par-

---

* You could concoct an energy landscape with big pits very low down and so trap the ball there. But, in nature, regions of low energy tend also to have small variations, whereas regions of high energy are prone to larger variations. This is a mathematical feature of "energy" itself, and it's why neural networks that can be shown to obey some energy function are more convincing as models of how the brain works than are neural networks that don't. The ones that do are "more natural."

ticular sort are all connected together, both weakly and sparsely, and at the same time if they are driven by some external oscillating input, then such a sparsely connected net of oscillators exactly duplicates the dynamics and capacities of a densely connected Hopfield net. When such a setup is modeled mathematically and then optimized to have the greatest possible capacity, the resulting structure corresponds exactly to the set of variable microcircuit structures found in the human cortex, including even the external influences from the other lower brains that impinge upon them.[8]

To the surprise of many researchers, both neuroscientists and computer scientists, these sorts of networks turn out to have the additional capacity to process changing streams of data in real time—just as the human fourth brain does. Indeed, such networks of oscillators open up an even deeper understanding of how the internal components that go into all cells—neurons included—function as small-scale analogs of networks. We can begin to consider all of life, at every scale, as inherently computational.

7

THE GAME OF LIFE, OR HOW THE LEOPARD GETS ITS SPOTS

What a game of chance human life is!
—Voltaire, *Les Délices* (November, 24, 1755)

When McCulloch and Pitts published their paper on neuronal learning, the great mathematician, quantum theorist, and computer scientist—the "Father of the Computer"—John von Neumann realized that the secret of spontaneous learning in nature had been cracked. He was therefore a consistent supporter of the Perceptron. But von Neumann glimpsed in the capacity for spontaneous learning something that went beyond computation and neuroscience. He was convinced that therein lay the secret of life itself. To uncover that secret was a Faustian goal, to be sure—but von Neumann was a man in the Faustian mode, and as it turned out, he was not being overambitious.

Von Neumann's intellect was evident in childhood. Since the age of six, Jancsi, as he was called, had been exchanging jokes with his father, a prominent banker, in classical Greek as well as in German and French. At gatherings at the Neumann family home in pre–World War I Hungary, a guest would select at random a page and column of the Budapest phone book, little Jancsi Neumann would glance quickly at the page, and then he would answer any question about it: names, numbers, addresses—or he'd simply recite the page verbatim. By the age of eight, he had already taught himself calculus.

While a student at the Lutheran Gymnasium, Jancsi came to the attention of an outstanding and devoted mathematical pedagogue, Lázló Rátz. In spite of the Neumann family's wealth, Rátz refused payment for private tutoring sessions. While still at the gymnasium, von Neumann (as he began to call himself justifiably, though his father chose not to use the honorific) published his first paper in mathematics. In Rátz's view, the honor of playing some part in the education of a child he knew was destined to become one of century's greatest mathematicians was more than sufficient recompense.

But Jancsi's family belonged to an elite circle of Jewish families who after the Enlightenment and the French Revolution had abandoned Judaism to rise rapidly in the surrounding, nominally Christian world, and Max Neumann was determined that his son should not return to what he saw as a merely secularized variant of the cloistered, highly intellectual existence of their religious forebears. He should pursue a career consistent with the family's prominent role in society. Jancsi was persuaded to strike a compromise, perhaps in part because he himself loved material things. So, after completing preparatory education, he earned a diploma in chemical engineering from the Eidgenössiche Technische Hochschule in Zürich, Europe's equivalent of today's MIT.

At the ETH, von Neumann studied no math, per his father's wishes. But nevertheless he sat for the mathematics Ph.D. qualifying examinations. He obtained one of the highest marks ever recorded and enrolled at the University of Budapest. George Polya, a mathematician of the highest caliber, was von Neumann's instructor: "Johnny was the only student I was ever afraid of. If in the course of a lecture I stated an unsolved problem, the chances were he'd come to me as soon as the lecture was over, with the complete solution in a few scribbles on a slip of paper."[1]

Von Neumann was soon invited to a professorship at Princeton. Shortly thereafter he was asked, with Albert Einstein, to become one of the first six mathematics professors at the newly established Institute for Advanced Studies. He was in no way overshadowed by Einstein. Eugene Wigner, himself a Nobel Prize–winning physicist described as one of the greatest of our time, had known Jancsi since childhood:

> I have known a great many intelligent people in my life. I knew Max Planck, Max von Laue, and Werner Heisenberg. Paul Dirac was my brother-in-law; Leo Szilard and Edward Teller have been among my closest friends; and Albert Einstein was a good friend, too. And I have known many of the brightest younger scientists. But none of them had a mind as quick and acute as Jancsi von Neumann. I have often remarked this in the presence of those men, and none ever disputed me.[2]

Nonetheless, von Neumann was hardly an ascetic: "Parties and nightlife held a special appeal for von Neumann. While teaching in Germany, von Neumann had been a denizen of the Cabaret-era Berlin nightlife circuit."[3] The parties he and his wife threw at their home in Princeton were famous, with the atmosphere of a European salon, at which Jancsi happily played the role he had learned as a child. On one such occasion someone posed to him the fly-and-train problem: Two trains twenty miles apart approach each other on the same track, both traveling at twenty miles per hour. A fly at the front of one train flies at sixty miles per hour to the second train, then immediately returns, and sets off again, over and over. How far does the fly travel before being squashed by the collision of the two trains? Von Neumann answered at once. Some among the assembled guests assumed that he must have discovered some simple trick that laid the solution open. Asked what it was, he responded, "No, I simply summed the series." Von Neumann, known for his keen sense of humor, was adding not the series, I suspect, but to his legend. There is, in fact, a trick, found by my wife, in a flash. The trains meet at the midpoint, which means they travel ten miles. The fly travels three times faster than the trains during the same amount of time, hence, thirty miles. This is the same answer you get if you break the trip apart into legs so as to form a converging infinite series. Then each leg covers three-quarters of the remaining distance, and the distance is halved each leg. This equals 3/4(1 + 1/2 + 1/4 + . . .) = 3/4(1 + 1) = 3/4(2)(20) = 30.

By the end of his brief life—he died in his fifties—there was hardly an important area of mathematics to which von Neumann had not made major contributions. His ideas generated seven Nobel Prizes, the most recent in 1997, in economics. He covered physics as well. His *Mathematical Foundations of Quantum Mechanics,* published in 1932,[4] established a whole new way of approaching the subject and remains a substantial part of its foundation to this day: "Quantum mechanics was very fortunate indeed to attract, in the very first years after its discovery in 1925, the interest of a mathematical genius of von Neumann's stature. As a result, the mathematical framework of the theory was developed and the formal aspects of its entirely novel rules of interpretation were analyzed by one single man in two years (1927–1929)."[5]

Von Neumann was also fascinated by games and almost single-handedly developed game theory. His ideas found immediate application in fields as diverse as economics and military strategy. The cold war doctrine of "mutual assured destruction" was directly based on his game theory, and it is a cruel injustice that filmmaker Stanley Kubrick modeled Dr. Strangelove—who loved war, not its prevention—after von Neumann.

"EVERYTHING FLOWS" (HERACLITUS)

It was inevitable that von Neumann would eventually tackle the pedestrian-sounding subject of turbulent flow—because for centuries it had beaten back all attempts at understanding. The last substantive progress had been made by Blaise Pascal in the late 1600s. To tackle the problem, von Neumann developed an entirely new approach—an approach with profound ramifications that are spreading still today and that lie close to the heart of our own subject.

Instead of trying to figure out mathematical solutions to the horrendously complex equations that describe turbulence, he devised a method of iterative approximation—guess, measure the error, adjust the answer, feed it back in again. It's not surprising that he would later so quickly grasp the power of neural network computation, since it, too, is fundamentally iterative.

The results of von Neumann's iterative calculations were never perfect but rather were precise to whatever degree desired. Not only did the method work for turbulence, it proved the only way of hurtling the major obstacle in the Manhattan Project. To perform the huge number of calculations needed to arrive at sufficiently precise values in the equations that allowed the minute portion of explosive isotope of uranium to be separated from the nonexplosive isotope in which it was mixed, von Neumann invented the modern computer. The computer not only made the bomb possible but was key to the Allies' cryptographic superiority—even more crucial to their victory than the bomb. And out of all that arose the National Security Agency as the world's most powerful intelligence agency. Were it not for von Neumann, it's doubtful that people like Stanley Kubrick would have the freedom to lampoon people like von Neumann. In fact, every computer made today, from the great supercomputers developed by the Department of Defense to the humble chip inside your daughter's Furbey, is based on what has come to be called "the von Neumann architecture."

"Guess, measure the error, adjust the answer, feed it back in again": This is the basic method of neural networks, which is to say of natural intelligence. In typically clear and startlingly deft fashion, von Neumann had actually developed the principles of parallel computation. That his ideas arose from the study of fluid dynamics has an important implication that cannot be seen without a radical change in perspective as to what nature is all about.

The mere swirl of a teaspoon generates whirlpools and ripples whose equations of motion are utterly beyond solution. But try to see it this way: Natural systems always assume their proper form. Instead of thinking, as is natural, "things just happen," imagine for a moment that nature is "modeling" something and to do this she must make a series of "calculations" of some kind, in a "computer" of some evidently material substance. In that case, she performs her "calculations" effortlessly. How does she do it? Of what is her "computer" constructed, and how does it work? What is the "algorithm"?

Certain answers are obvious. For one, there is no "she": no orchestrator, no one who stands apart and models and calculates. Instead, there are only a myriad of individual parts—bits of matter (in the case of fluid turbulence, molecules of air or water, for example) following no instructions beyond the physical laws that govern their individual motions and immediate-neighbor interactions. Via the local interactions between parts, the "correct"—the only possible—pattern of global order emerges. A cup of coffee "computes," in this strange perspective, in the same sense that a spin glass "remembers."

So perhaps the best way to "solve" the problem of turbulence—"Given such and such conditions, what's the cup of coffee going to look like one, two, three seconds after I stir it?"—is to string together a bunch of simpleminded "processors" interacting with their neighbors. Each processor need do no more than respond to stimuli (inputs) with movements (outputs) as would a typical water molecule (very weak coffee). Neighbor-to-neighbor interactions between processors would be like the transfer functions in neurons that determine the output based on the total input. The "rules" of such a network would just mimic the known physical laws that govern intermolecular behavior.

No single such processor—von Neumann called them *cells*—nor the designer himself would or could analyze a problem in fluid dynamics to arrive at the solution—an equation. But in the way that nature does it, all the cells working at once, "in parallel," could "solve the problem" by producing the proper output for any given set of initial conditions. Von Neumann termed such a massively parallel processor composed of identical processing elements with given neighbor-to-neighbor interactions a "cellular automaton," or CA. Like God "the absent watchmaker," you set your cellular automaton going and come back whenever you want—if ever—to see what it's come up with. It's worth noting that von Neumann chose the word "cell" deliberately. His real goal, like ours, lay far beyond an understanding of eddies or the creation of a kind of computer that could make bombs: "Though von Neumann was a lead-

ing physicist as well as a mathematician . . . in his work on cellular automata
. . . his interest was directed more at a reductionistic explanation of certain aspects of biology."[6]

He developed cellular automata in a quest to discover *the minimally essential features of a system capable of the parallel computational complexity of
life itself.*

ENTER THE AUTOMATON

Figure 7–1a illustrates the basic form of von Neumann's cellular automaton.
Figure 7–1b shows a variant in more common use today—a so-called Moore
Neighborhood CA.

Every cell plus four (or, as on the right, eight) neighbors constitutes a CA
"neighborhood." Each cell is assigned an initial value. These may be as basic
as 0 or 1 (or –1 and +1), or any of hundreds of different values. Cells change
their value according to an "update rule" based on their own and their neighbors' values. For example, assume that cells may take on values of either 0 or
1. Then the so-called "majority rule" states that if the sum of the neighborhood values plus the cell's own value is 4 or less, then the cell will be updated
to 0; if greater than 4, then to 1. In practice, all cells are updated at once, based
on some present value of the cells. Then they are all updated at once again,
based on the values that the previous update produced. This kind of simultaneous updating is called "synchronized."

An update rule is very like the rule that determines what will come out of
a neuron given what comes into it; or like the laws that determine how spins
will align in the frustrated, mixed ferromagnetic and antiferromagnetic neighborhoods of a spin glass. Indeed, figure 7–1c shows that a CA is very like a locally connected Hopfield "map" made up of two triangular, overlapping "canonical microcircuits" (of the kind shown in figures 5–4, 5–5 and 6–6).

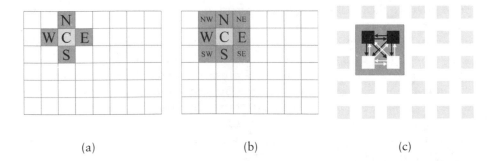

(a)   (b)   (c)

FIGURE 7–1   Schematic of a cellular automaton. (a) 5-cell "von Neumann neighborhood." (b) 9-cell "Moore neighborhood," a variant developed by one of von Neumann's followers. There are no limits as to what a CA neighborhood might look like—
if in the end it works. (c) A cellular automaton represented as a neural network composed of two triangular canonical microcircuits that are partially overlapping.

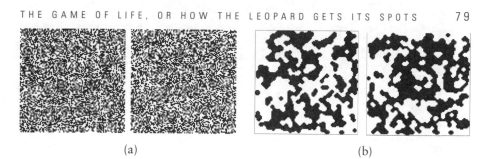

(a)   (b)

FIGURE 7-2   (a) Cellular automaton in two different randomized initial states and (b) their corresponding final states.

Like neural networks, CAs self-organize: they generate global order from purely local interactions. Figure 7–2a shows a 100×100 cellular automaton in two different randomized initial states of 0s and 1s, white equaling 0 and black 1. After only ten updates using the majority-vote rule, the CAs attain the respective final states in figure 7–2b.

Of all possible configurations of black and white squares into which a 100×100 board might arrange itself, it's clear that this is of a distinctive family. We might call it "Guernsey Cow," and we know immediately that it belongs to a different family from a pattern we were to call "Zebra" or "Polka Dot." The "family" to which a specific configuration belongs constitutes the solution, or energy minimum, typical for a given system—that is, an aggregate of cells with a specific update rule. There are many solutions—states of approximately equal minimum energy; patterns belonging to the same family—for any given system.

Nonetheless, the two final patterns above are distinct, and any given location in the original grid has about an equal likelihood of ending up black as white, regardless of what it starts out as. This is so even though the patterns are recognizably the "same" (in the way that Juliet "is" the sun)—just as the pattern of black and white on a Guernsey cow is distinctive yet differs in its specifics from animal to animal. In fact, the regular yet unpredictable colored patterns found on the skins of many animals—cows, tigers, zebras, fish—are generated in just this way—via local CA-like interactions among pigmented (biological) cells—as also are such infinitely variable yet strikingly similar patterns as our fingerprints.

Compare the cellular automaton in figure 7–2 to a simulation of spin-spin interactions, equivalent to a Hopfield-type neural network, also starting from two different, initially random configurations and modeling the actual physical laws found in nature that govern how nearby spins affect one another. (See figure 7–3.) Black represents spin-down and white spin-up.

Clearly, the process of self-organization in both systems is strikingly similar. Even more sophisticated cellular automata can be created in which both time and space are disordered: that is, the cells are not located in a nice array but distributed randomly over a surface (or in space); the cells are likewise updated at random. Furthermore, the update rule can be made "stochastic" instead of fixed—the equivalent of adding heat. But if the proper neighborhood

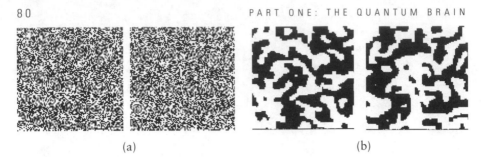

(a) (b)

FIGURE 7-3   Ising model for spin-spin interactions in a magnetic material. (a) Two different, initially random (disordered) configurations settle in (b) two different but similar final states.

rules are implemented, randomized initial states in these yet more lifelike, so-called amorphous CAs still produce order, indeed, an order that looks self-evidently more "natural," as shown in figure 7–4.

FIGURE 7-4   Amorphous cellular automaton generating a stable orderly distribution. Cells (nonblack) are arranged at random and may take on a range of values from white to dark gray.

Such automata come even closer to mimicking the "global processing characteristics of genuinely living systems." The jargon highlights the idea that computation ("processing") is something that nature (life especially) performs just by doing what it does, and that its kind of computation involves generating patterns ("global" order) not according to a top-down plan but solely depending on the rules that govern local relations among bits of matter. In short, figure 7–5 shows how a leopard gets its spots, as modeled by an amorphous cellular automaton.

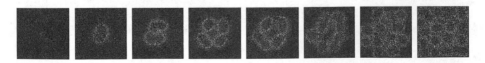

FIGURE 7-5   Amorphous cellular automaton generating "leopard's spots." The final pattern is stable.

LIFE: A UNIVERSAL COMPUTER

Like neural nets, CAs generate global spatial order through strictly local interactions. They perform logical operations through the use of oscillators, just as neural networks of a particular sort can. Using oscillation, can CAs also generate global temporal order, that is, patterns that change over time in an orderly way ("synchrony")? (The problem was first addressed in the late 1950s and early 1960s. Marvin Minsky was the first to prove that CAs were capable of synchronization, but his proof was never published.) Here's how they can do it:

1. If more than two of its eight nearest neighbors are "awake" (set to 1), a cell goes to or remains "asleep" (set to 0). This is a simple threshold form of inhibition.

2. If a cell falls asleep, it remains asleep for three updates, then automatically awakens.

Biological neurons have a "refractory" period. After one discharge, further firing is inhibited for a time. In its fully ready state, the neuron has a "hair trigger"—its threshold for discharge is very low, and it takes very little incoming excitation to set it off. Immediately upon firing the threshold jumps way up, to settle down only gradually. A constant excitatory stimulus will therefore cause a neuron to fire rhythmically at a rate determined by the intensity of the excitation. In this way, one continuous input meeting a timed refractory period results in a "tunable" oscillatory output—increase the rate by upping the input intensity. The situation with our new CA is rather similar, except that the refractory period is fixed. (The rule is easily modified to create a tunable cell.)

Tommasso Toffoli and Norman Margolus, pioneers in CA theory at MIT, compare this kind of CA to the life of tubeworms on the ocean floor,[7] a model of *competitive/cooperative interaction* in living systems. A tubeworm pokes out its tendril-covered head to sieve for floating bits of food. If it detects too many neighbors with their heads out, it ducks back in for a fixed period of time before tentatively emerging again. A colony of tubeworms thus sets up a dynamic pattern of "cooperation" within its natural competition for resources.

Figure 7–6a shows an initially random state of such a CA. It's more complex than the CA in figure 7–2 because each cell can take on a range of values (grays, instead of just black and white) and there are many more elements. As updating proceeds, global patterning progressively emerges. Two later states are shown in Figure 7–6b and c.

After roughly 250 updates, a high degree of organization is present. But this snapshot can't show how rather than converging on a stable pattern (as in figures 7–2, 7–3, 7–4, and 7–5), this CA converges on a dynamically changing, roughly *cyclical sequence* of traveling waves and spirals that radiate out from various centers that *never* repeat exactly yet are unmistakably of a stable "family." Some impression of this incredibly well-organized global dynamic may be gleaned from the sequence of snapshots set side by side in figure 7–7.

(a)   (b)   (c)

FIGURE 7-6   (a) Initial randomized state of a compete/cooperate cellular automaton with "refractory" neighborhood inhibition. (b) Compete/cooperate cellular automaton after 50 updates. (c) Compete/cooperate cellular automaton after 254 updates.

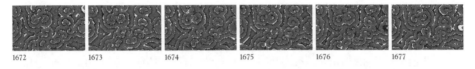

1672   1673   1674   1675   1676   1677

FIGURE 7-7   Synchronized traveling concentric waves and spirals in a compete/ cooperate cellular automaton with "refractory inhibition." Updates 1672–1677.

Such global oscillations mean that the entire system has become synchronized. The "refractory period" plays the role of a local timekeeper, as it does in neurons. No one says "Synchronize your watches," yet they do so anyway. Toffoli and Margolus have likewise shown that even if the updating is random, oscillatory CAs will still self-organize both spatially and temporally and that all of the basic logical functions, including nested XORs, can be built up from them. They are, in short, parallel computers in exactly the same way that neural networks of the unsupervised sort are.[8] More recently, a DARPA-funded project at MIT has shown how synchronized waves of a very similar sort may be generated in CAs that not only lack synchronized updating, but even their cells can be scattered at random.[9]

HONEY, I SHRUNK THE COMPUTER

Von Neumann saw that, for the living organism, the cell is the fundamental local processing element that generates global patterns and global intelligence. But single-celled organisms demonstrate both orderly patterns in their behavior and rudimentary intelligence. If the basic idea behind neural networks and cellular automata is correct—that intelligence and order at one scale arise bottom-up from the local interaction of many elements at a smaller scale—then the individual cell itself (the neuron in particular) must contain yet smaller elements which, interacting locally, can likewise generate intelligence and global order. Is there any evidence for such a thing?

Yes. In the early years of the twentieth century, an earnest Russian teenage boy set out to build a bomb to kill the czar. He didn't succeed, but his investigations led Boris P. Belousov to an unusually successful career as a military chemist in the new Soviet Union. During his service, he stumbled upon a phenomenon never seen before: an *oscillatory* chemical reaction—its color rotated among hues at regular time intervals, with the change—hence the reaction—propagating as a wave front. Belousov submitted his findings to a Soviet journal of chemistry.

The paper was rejected out of hand because the phenomenon in question was flatly impossible. By the laws of thermodynamics, the editor replied haughtily, chemical reactions in a closed system must move swiftly toward their equilibrium state. Furthermore, they move in but one direction—two steps forward, one step back was out of the question.

Belousov was so disgusted that he quit science altogether, despite the entreaties of his colleagues. In the mid-1950's he turned the recipe for his reaction over to a biochemist, Sergei Shnoll, who went on to become professor at the Institute for Theoretical and Experimental Biophysics in Pushkin. The institute published Belousov's article, and it came to the attention of a physicist, A. M. Zhabotinsky, who studied it carefully. By 1966 the first world congress on the Belousov-Zhabotinsky reaction and other oscillating and "excitable" chemical systems was held. It is now a major area of study among chemists. Figure 7–8a shows what the Belousov-Zhabotinsky reaction looks like when it is carried out in a thin layer under carefully controlled conditions. To make the comparison easy, figure 7–8b reproduces an enlarged version of the temporally coherent CA of figure 7–7, after it has settled into a stable pattern of oscillatory waves and spirals. The two dynamic patterns are of an identical form, except that the chemical reaction is vastly more "fine-grained." Its "cells" are molecules, and there are trillions of them. The 100×100 CA has a mere 10,000 cells.

In the words of one researcher: "[D]igital logic can be implemented in the chemical kinetics of homogeneous solutions: We explicitly construct logic gates and show that arbitrarily large circuits can be made from them. This proves that a subset of the constructions available to life has universal (Turing) computational power."[10]

The same underlying computational processes evidently govern self-organization at many different scales in nature. Figure 7–8c is a photograph of the growth and migration of a colony of a particular amoeba, *Dicytostelium*

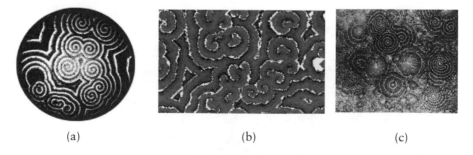

(a)   (b)   (c)

FIGURE 7–8   (a) Belousov-Zhabotinsky oscillatory chemical reaction. (b) A compete/cooperate cellular automaton after it has developed temporal coherence, shown at update 1677. (c) Growth pattern of *Dicytostelium discoideum*.

discoideum, for which the individual processing element, or cell, is in fact a cell—a single-celled organism. Of course, the waves and spirals of the amoeba colony travel very much more slowly than those of a chemical reaction or in an arbitrarily fast artificial parallel processor such as a CA; it may take years for the amoebas to stabilize such a pattern.

COMPUTATIONAL EMBRYOGENESIS

One of the great puzzles of life is how its various forms take shape. For centuries the intricacies of the body's form, and its development from a mere blob of cells, has also served as a powerful argument in favor of a designer who sculpts these forms according to his will. Can amorphous cellular automata of the sort that, in two dimensions, give the leopard its spots through purely local interactions really generate far more complex three-dimensional forms that themselves go on to compute? We have already raised some reasonable doubts as to the purely *self*-organizing nature of artificial neural networks. For even if once established they do indeed learn by themselves, we design their architecture, and their processing schemes, and we assemble the thing to make it work (or write the code that allows a standard computer to perform a software simulation of it). In other words, it's one thing to have shown that the brain can learn by itself, but can a self-organiz*ing* brain itself be self-organiz*ed*? How does such a brain first come into existence?

Figure 7–9a is a famous early drawing (from 1940) of the first significant stage in the development of the brain: the formation of a hollow tube within a dense ball of cells. It is from this "neural tube" that all four brains will ultimately develop.

Compare this to the sequence of a three-dimensional amorphous cellular automaton (shown in two-dimensional cross-section in figure 7–9b) generating a neural tube–like structure strictly via local interactions among neighboring cells. (In living embryology, the local interactions are in fact mediated chiefly

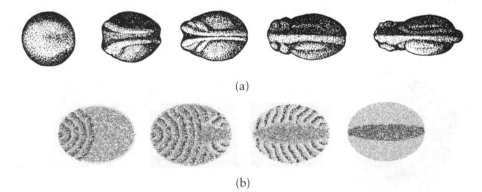

(a)

(b)

FIGURE 7–9   (a) Neural tube development from 50 to 80 hours postfertilization in embryonic *Rana pipiens* (frog). (b) Three-dimensional amorphous cellular automaton shown in two dimensions, performing neural tube embryogenesis via local interactions.

by chemical gradients.) The method is directly modeled on the "parallel computations" carried out in oscillatory chemical reactions.

There is an even more direct connection, however, between the self-organization of the embryo and the formation of the fourth brain as a functioning neural network. Ronen Segev and Eshel Ben-Jacob, the authors of a recent article in the *Journal of Complex Systems,* write:

> During embryonic morphogenesis, a collection of individual neurons turns into a functioning network with unique capabilities. Only recently has this most staggering example of emergent process in the natural world begun to be studied. Here we . . . propose that the embryonic environment (the neurons and the glia cells) acts as an excitable media [i.e., a cellular automaton] in which concentric and spiral chemical waves provide a mechanism for communication, regulation, and control required for the adaptive self-wiring of neurons.[11]

In other words, during the development of the embryo, amorphous cellular automaton-like processes serve as a "medium" out of which self-organizes the higher-level (and larger scale), neural network architecture of the brain, which itself then self-organizes the yet higher-level (and larger-scale) data of experience. Furthermore, it is specifically "chemical waves" at the next lower level (and smaller scale) that provide the communication medium for the embryonic cells. The arrangement strongly suggests that chemical systems of a certain sort may themselves function as CAs—or, if you wish, neural networks. And why not? If inorganic magnetic materials have the wherewithal to organize themselves, why not molecules and biomolecules? And if chemicals can do it between cells, as in embryogenesis, they surely can do it within each cell, within each neuron in particular, especially since the actual chemical signals and chemical reactions both within and between cells are the same.

JOURNEYING DOWN

Segev, cited above, was quite specific about the kinds of "chemical waves" involved in self-organization: They are *traveling* and *spiral* and *concentric,* just as in colonies of amoebas, in the Zhabotinsky chemical reaction, and in synchronous cellular automata. The significance of these characteristics is surprisingly simple in light of our discussion of modified Hopfield neural nets: Waves are oscillations; spiral and concentric waves are two- or three-dimensional stable states; networked oscillations can compute because they can form the mathematical equivalent of very efficient Hopfield nets; traveling oscillations can communicate signals and demarcate time so as to establish temporal coherence—synchrony. Chemical systems of this sort show every indication, therefore, of being capable of the same self-organizing intelligence as are neurons.

It all sounds very like our speculations about nested sets of HERs, each larger set organizing the smaller set within—*but in reverse.* In life, it is the interior that makes available to individual elements just those features they need so that the element, and its local neighbors, self-organize global structure. This kind of bottom-up self-organization is a form of parallel computation. In dropping down to the cellular level (neurons), we should therefore strongly suspect that we have by no means reached bottom.

In fact, it has long been known that the individual neuron is itself an extraordinarily complex entity. It alters its extrusions so as to create, modify, and destroy connections with other neurons; it is capable of self-regeneration, and collections of neurons can even increase their number in response to challenging mental tasks, for instance.[12] (That neurons do regenerate is a relatively recent finding, and overthrows long-established conventional wisdom.)

Clearly the interior life of the neuron goes far beyond a simple "transfer function" that adds up inputs. Given all that we have learned, we should suspect that this interior life is one of yet smaller-scale yet *more* massively parallel computation of some sort. It may not *look* like computation at first glance (just as the brain's method of operation didn't)—and who would ever look at a colony of amoebas and think, "Aha, computers!"—but studied properly, it's going to reveal itself as a variant of a Hopfield net, or a cellular automaton, or a net of oscillators, and they all compute in pretty much the same way: bottom up. *Each individual neuron, like each individual in society, is itself likely to be a network composed of many individual processing elements.* Different physical implementations of parallel processing predominate at different scales, but all implementations share a common computational and self-organizing capacity.

How far downward may we expect this parallelism to go? Edward Fredkin, one of the great early artificial intelligence pioneers at MIT, concluded that parallel computational capacity is woven into the very fabric of matter itself; that reality is, as it were, a massive cellular automaton, and that intelligence at every level was a necessary concomitant. In this perspective, turbulent flow really *is* computation. If such a notion makes sense, then we ought to find that the "design process" which eventually yielded a nervous system capable of functioning as highly as ours (and other systems too) likewise involves some kind of self-organizing, parallel computation.

That is precisely how we ought to consider *evolution.*

For those of you who still hang on fondly to your old Gilbert chemistry set, here is the formula for the Belousov-Zhabotinsky reaction:

- Equipment
  100 g balance, 25 ml graduated cylinder, 100 ml graduated cylinder, 250 ml graduated cylinder, magnetic stirring hot plate, 1500 ml Pyrex beaker, 1000 ml Pyrex beaker, four 500 ml Pyrex beakers, 250 ml Pyrex beaker, petri dish
- Solution (1): 800 ml water, 60 gm $NaBrO_3$, 24 ml concentrated H_2SO_4
- Solution (2): 25 g Malonic acid, 250 ml water
- Solution (3): 0.5 g o-phenanthroline, 1.0 g hydrated ferrous ammonium sulfate, 100 ml water
- Procedure: Mix together 180 ml Solution A, 130 ml water, 30 ml Solution B, and 10 ml Solution C
- . . . and stir continuously. Results:

A red solution that, at 25°C, flashes blue every ca. 18 seconds for approximately 1 hour and 15 min.

8

BY YOUR BOOTSTRAPS

Evolution is what it is. The upper classes have always died out;
it's one of the most charming things about them.
—Germaine Greer

Here's where we have come: From the computational perspective, physical reality is inherently like a cellular automaton, and thus facilitates computation and self-organization at all scales. The implementation of basic brain structure—neural embryogenesis—occurs via cellular automata-like processes. The huge intelligence of the matured and educated brain is largely the result of similarly self-organized learning within the *neural network* that was created during implementation. From a mathematical point of view, neural networks and cellular automata are almost identical—only the physical basis and appearance are different: Self-organization at one scale yields the capacity for self-organization at the next.

To this we add the following point: Via the effect of competitive breeding on a *genetic blueprint*, the design of the brain evolved—or, rather, *evolves*. The self-organization also takes place across huge time scales, with every biological generation an "update." Why "blueprint"? Before either DNA or RNA had even been discovered, John von Neumann deduced from first principles that life must be a kind of cellular automaton and that *a cellular automaton capable of reproducing itself,* hence of generating ever-improved versions of itself, *would need to contain its own blueprint, would need to have a means of copying that blueprint,* and *would need to have a means of constructing a new automaton according to that blueprint.* To improve the automaton, the *blueprint* would first have to undergo change. This is precisely the mechanism biologists later discovered to lie at the heart of reproduction and evolution.

Self-organization of the design *of the brain takes place at the level of the blueprint.* When the blueprint for a cellular automaton is altered, randomly, it leads to an altered version of the entity when next it duplicates itself from the blueprint. If the altered version is superior—that is, better adapted to its conditions—it should preserve the modified blueprint, discard or in some other fashion inactivate the old, and continue the modification process—like HER losing her jellybeans.

GENETIC ALGORITHMS: SYNCOPATED MATING

Suppose we have implemented some kind of *self-replicating cellular automaton*, along the lines of one developed in 1984 by Christopher Langton, now director of the artificial life program at the Santa Fe Institute. Figure 8–1 shows it at three stages.

In time, a single Q-shaped Langton CA becomes an expanding "self-similar" Q-shaped *colony* consisting of many versions of itself surrounding a growing core of dead skeletal O-shaped remains.

This simple CA shows many features of a living colony of bacteria or fungi. However, even though it is self-replicating, it is not *adaptive*. The rules that have been embedded in the cells that allow it to generate copies of itself in unoccupied cells are permanently fixed. So, let's give it the capacity to adapt, by embedding within it von Neumann's blueprints.

First, we represent the update rule in a code: a string of numbers (0 or 1, for simplicity) that we call the "X-chromosome." Let's say that twenty digits are sufficient to encode the rule. Then each of its twenty components are "genes" as, for example, in table 8–1.

Since each gene may take on one of two values, the total number of possible chromosomes is 2^{20} = 1,048,576. Suppose we now wanted to find the "best" rule (where "best" produces some desirable result, whatever that is). We could search for it by seeing what results we get when X = 00000000000000000001, 00000000000000000010, . . . , all the way to 11111111111111111111—over a million trials. Or we could choose rules at random on the (perhaps mistaken) assumption that rules numerically "close" to one another will not demonstrate terribly different capacities.

These optimization methods will be highly inefficient at finding the best chromosome. Here's another approach: We select at random 100 different X chromosomes and test them all on our problem. We rank order the results, creating a "fitness" hierarchy among them. We make multiple copies of the best five *and pair them at random*.

FIGURE 8–1   Langton self-replicating cellular automaton at 0, 56, and 157 updates, respectively, from left to right.

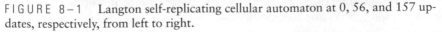

TABLE 8.1   A synthetic "chromosome" consisting of twenty binary "genes."

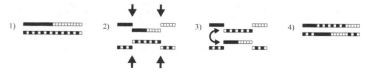
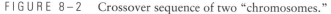
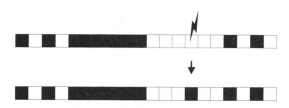
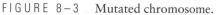

FIGURE 8–2   Crossover sequence of two "chromosomes."

FIGURE 8–3   Mutated chromosome.

We line up next to its partner each paired chromosome and pick two spots at random along their lengths. We cut both chromosomes at these two points, extract the segment created by the cuts, and *switch them*. Then we reconnect the ends, thus creating two new but related chromosomes, as in figure 8–2.

This process, referred to as "crossover," is modeled directly on real biological events during *reproduction* that are a crucial part of *evolution*. Since we have multiple copies of every pairing, we may allow each pair to undergo many different crossover schemes. Two different, imperfect but good solutions may each contain parts of a better solution. By cutting the solutions up and recombining them, we push the search in the direction of what is *probably* (though not necessarily) an improvement.

The mating dance so far yields only scrambled parents. To ensure that some descendants will explore novel solutions, we "syncopate" the algorithm—we give it an unexpected jog in the middle of its regular rhythms: Newly crossed-over chromosomes are deliberately made to risk that one or more of their genes will suddenly change at random. (See figure 8–3.)

All the new chromosomes compete against each other and their parents and the whole process is repeated. As the generations unfold, better and better chromosomes emerge. In this way, perhaps we can create an evolving, self-improving cellular automaton.

A MILLION MONKEYS

For example, consider a 25-gene chromosome each of whose genes has 101 possible variants. Each gene codes for a lower- or upper-case keystroke of an extended PC keyboard. As its creator, we decide that the most fit chromosome is as shown in table 8–2.

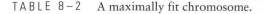

TABLE 8–2   A maximally fit chromosome.

v	g	B	E	l	X	x	+	3	w	K	f	c	f	R	K	_	7	9	N	,	E	8	h	/

TABLE 8.3   A chromosome composed of a random selection of genes.

What are the odds of obtaining this sequence merely by chance? There are $25^{101} \cong 10^{50}$ possible combinations. If we set a million monkeys at typewriters, each generating one sequence per minute with never a duplication, the correct sequence should appear in about 10^{44} years—a trillion trillion trillion times the age of the universe.

Compare this to a genetic algorithm turned loose on the monkeys. We start with a completely random selection of genes. (See table 8–3.)

We define the fitness of a sequence as the sum of the distances of each character (on a keyboard) from the correct one, doubled if the wrong case, and scaled so the initial sequence has a fitness rank of 1,000 and a perfect sequence has a fitness rank of 0. Set the maximum crossover at 5 percent and the mutation rate at 7 percent. After two generations of competition, mating, crossover, and mutation, the best chromosome is shown in table 8–4.

No genes are correct, but its fitness value is 486, already halfway there! The sixteenth generation produces the chromosome of table 8–5 with three correct genes and a fitness of 235.

Convergence to the ideal chromosome (a "basin of attraction") follows the sequence laid out in table 8–6.

Generation 91 produces the perfect chromosome. Fitness (the inverse of "error") progresses much as convergence does in neural nets, CAs, or HER, as the graph in figure 8–4 illustrates.

Evolution in a more complicated, two-dimensional fitness (energy) landscape looks even more similar to convergence in Hopfield networks. The fitness landscape in figure 8–5 is shown in "topographic" form. Each circle represents a chromosome; its location on the landscape represents where it is with respect to coding for the optimal features for this environment. As evolution proceeds, better and better chromosomes emerge (better meaning more fit for this landscape), and the poorer ones disappear. There are altogether fewer and fewer chromosomes, since while there are many ways for different chromosomes to be equally *mal*adaptive, there is only one best adaptation. The remaining adaptive chromosomes therefore *converge* on a basin of attraction, growing ever more alike.

L	5	F	d	X	w	l	y	a	0	l	.	%	t	u	D	#	E	%	b	#	+	Y	o	+

TABLE 8–4   A chromosome after two generations.

h	n	a	k	/	c	e	w	l	P	o	r	*	s	I	g	,	e	*	c	e	#	+	r	7

TABLE 8–5   A chromosome after sixteen generations.

Row	1	2	3	4	5	6	7	8	9	10	11	12	13	14	15	16	17	18	19	20	21	22	23	24	25
51	H	n	n	j	c	v	S	z	v	r	o	t	e		S	h	;	i	b	k	m	e	a	r	e
52	H	n	n	j	c	v	S	z	v	r	o	t	e		S	h	;	i	d	p	m	e	a	r	e
53	H	n	n	j	c	v	s	z	v	r	o	t	e		S	h	;	i	d	p	m	e	a	r	e
54	H	n	n	j	c	v	s	z	v	r	o	t	e		S	h	a	i	b	r	o	c	a	r	e
55	H	n	n	j	c	v	s	z	v	r	o	t	e		S	h	;	i	d	r	o	e	a	r	e
56	H	n	n	j	c	v	s	z	v	r	o	t	e		S	h	a	i	d	r	o	e	a	r	e
57	H	n	n	j	c	v	s	z	w	r	o	t	e		S	h	a	i	d	r	o	e	a	r	e
58	H	n	n	j	c	v	s	z	w	r	o	t	e		S	h	a	i	d	r	o	e	a	r	e
59	M	n	n	j	e	v	s	z	w	r	o	t	e		S	h	a	i	d	r	o	e	a	r	e
60	M	n	n	j	e	v	s	z	w	r	o	t	e		S	h	a	i	d	r	o	e	a	r	e
61	M	n	n	j	e	v	s		w	r	o	t	e		S	h	a	i	d	r	o	e	a	r	e
62	M	n	n	j	e	v	s		w	r	o	t	e		S	h	a	i	d	r	o	e	a	r	e
63	M	n	n	j	e	y	s		w	r	o	t	e		S	h	a	i	d	r	o	e	a	r	e
64	M	n	n	j	e	y	s		w	r	o	t	e		S	h	a	i	d	r	o	e	a	r	e
65	M	n	n	j	e	y	s		w	r	o	t	e		S	h	a	k	d	r	o	e	a	r	e
66	M	n	n	j	e	y	s		w	r	o	t	e		S	h	a	k	d	r	o	e	a	r	e
67	M	n	n	j	e	y	s		w	r	o	t	e		S	h	a	k	d	r	o	e	a	r	e
68	M	n	n	k	e	y	s		w	r	o	t	e		S	h	a	k	d	r	o	e	a	r	e
69	M	n	n	k	e	y	s		w	r	o	t	e		S	h	a	k	d	r	o	e	a	r	e
70	M	n	n	k	e	y	s		w	r	o	t	e		S	h	a	k	d	r	p	e	a	r	e
↓	↓	↓	↓	↓	↓	↓	↓	↓	↓	↓	↓	↓	↓	↓	↓	↓	↓	↓	↓	↓	↓	↓	↓	↓	↓
90	M	o	n	k	e	y	s		w	r	o	t	e		S	h	a	k	d	s	s	e	a	r	e
91	M	o	n	k	e	y	s		w	r	o	t	e		S	h	a	k	e	s	p	e	a	r	e

TABLE 8.6   Convergence to the perfect chromosome.

Recently evolutionary biologists have found situations in nature where organisms whose ancestors were very different evolved convergently, as if certain descendants of both birds and fish were to be very similar mammals—a vastly exaggerated analogy, however. Such examples are rare in nature because

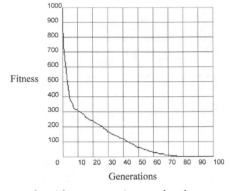

FIGURE 8–4   Genetic algorithm converging on the chromosome whose genes spell "Monkeys wrote Shakespeare."

the fitness landscapes are extraordinarily complex and "rugged," as it's called, with many deep but still high-energy relative minima (volcano craters high in the mountains) and many equally low energy minima (valleys) near the absolute minimum. Many different species with wildly different features can adapt equally well to the same environment—which environment, incidentally, includes each species for every other! The two dimensions in figure 8–5 imply an environment with only two critical characteristics. Natural environments have countless thousands of characteristics and would require a diagram with countless thousands of perpendicular dimensions.

Neural networks, cellular automata, genetic algorithms, and simulated annealing are all methods that perform massively parallel computation in the way that nature does. They all are closely related to one another mathematically. Which form nature chooses in any given situation—or which combination of methods—is determined by the materials at hand with which it evolves the most efficient angle of attack. The same principle applies in artificial applications. CA-based massively parallel machines do a particularly good job of simulating environments such as those within which evolution "computes" its solutions.

But unless you restrict your attention to one or two variables at a time, the results can scarcely be represented visually. Nonetheless, the numerical results show the same type of convergence as in "toy" examples with a few di-

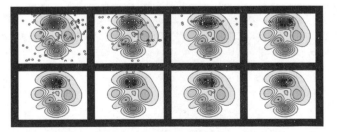

FIGURE 8–5   Genetic algorithm in a two-dimensional fitness landscape with numerous secondary basins.

mensions. They also show the large-scale, long-time oscillations that are the hallmark of networked, iterative, self-organizing computational process. So, for example, evolutionists now take seriously the possibility that the rise and fall in the numbers of a species, and even ebbs and flow in the emergence and disappearance and reappearance of species of a certain type (e.g., carnivores), may have little to do with changing external circumstances and arise, rather, because of the natural dynamics of self-organizing processes. Mankind may turn out to be but the momentary blue flash in the background red of an eons-long biospherical equivalent to the Belousov-Zhabotinsky reaction.

FROM GENETIC PROGRAMMING TO MOLECULAR COMPUTING: EVOLUTIONARY HARDWARE

Whether alone or in combination, artificial parallel computational methods are most frequently implemented in software. Lines of computer code represent the mathematical functions that correspond to "spin" and to spin-spin interactions; to a "cell" and its cell-to-cell rules; to a "neuron" and its connections; to "genes" on a "chromosome" and its mutations, crossover, and offspring. With remarkable success, many groups around the world, in both industry and academia, are developing wholly evolved, continuously evolving software.[2]

When we run genetic algorithm, neural network, cellular automata, or simulated annealing programs, we are computing via *software* running on *fixed hardware.* This distinction does not exist in nature. When the swirl of water about a teaspoon "solves" the hydrodynamic equations that stumped Pascal, or when the synapses in the brain increase or decrease their strength in response to training, there is no "program." From reproduction at the level of chromosomes to evolution of a biosystem, the hardware modifies (organizes) itself and the pattern of modification constitutes the only "programming."

In the realm of artificial intelligence, genetic algorithms evolve *blueprints* for a better neural network; we manually convert the best into a software implementation. How about an artificial version of "evolutionary molecular computation," in which genetic algorithms directly shape *hardware,* without us middlemen? That would be much closer to nature—no humans interpreting and constructing. Surely such a scheme sounds like the literary fantasies of a Mary Shelley or an Isaac Asimov. But, in fact, early prototypes of evolving, self-organizing hardware have already been built. A *fifth* international congress devoted to hardware evolution is under way as I write.

Here's how it works. Let's imagine the ultimate in electronic hobby kits. Instead of a pegboard into which we fit various elements and wire them together as in figure 8–6a, we have an array of identical elements, each of which you can switch to be a capacitor, a resistor, a transistor, a segment of wire, and so on as you choose, as in figure 8–6b. To create any device, we set all the switches to form the proper circuitry.

We needn't set the switches by hand, either; we use software-driven controllers.

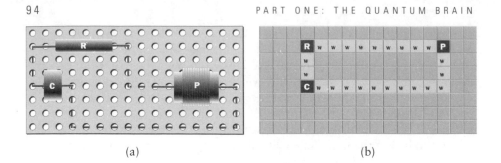

(a)   (b)

FIGURE 8-6   Modularized circuits. (a) Electronics kit implementing a voltage (P), resistance (R), capacitance (C) circuit. (b) Regular array of switchable elements implementing the same circuit.

Such devices now exist, called "field programmable gate arrays (FPGAs)." To design a processing chip (instead of a radio), we just breed (in the driving software—the blueprint) a large number of randomized FPGAs set to implement *the logical operations that all computers use*, any of which can be created by the correct arrangement of resistors and capacitors and, in particular, transistors.

These operations, called "logic gates," take an input and generate an output. Thus, a NOT gate takes a single input and puts out a single output. If the input is 1, NOT puts out a 0; if the input is 0, NOT puts out a 1. 1 can be represented by the presence of current, 0 by its absence. An XOR gate takes *two* inputs and puts out one. If the input is 1,1 or 0,0, XOR puts out 0; if the input is 0,1 or 1,0, XOR puts out 1. Other gates are AND and OR. Any set of computer functions can be built up out of a series of such gates linked together.

In a FPGA, logic gates can be switched in *nanoseconds* (10^{-9} seconds). If the software running the genetic algorithm were itself implemented in dedicated hardware, an evolutionary process that requires 100,000 generations can occur in under one second! John Koza at Stanford anticipates that with FPGAs "you could . . . develop brain-like self-learning."[3]

In the first working test case, Adrian Thompson and Philip Husbands at the Centre for Computational Neuroscience and Robotics at the University of Sussex bred an FPGA to discriminate between two musical tones. Now, electronic circuits don't have ears. To an FPGA a "musical tone" is a frequency of oscillation, and a frequency is a measure of time: wave peaks per second. The FPGA therefore had to evolve a method of discriminating time data, even though neither clock nor oscillator was built into the array—just logic gates. Furthermore, "We were asking incredibly fast components to do something much slower, and produce very nice, steady output," explains Thompson.[4] "I wanted to see what happens if you let evolution break out of the constraints that humans have. If you give it some hardware, does it do new things?"[5]

After 4,100 generations, the system evolved an ability to discriminate perfectly between the two tones. But when Thompson and associates "opened the black box" to see how it did it, they were surprised indeed. Their FPGA had one hundred logic gates. When they traced the evolved connections, they found that only thirty-two of them were connected at all. Yet if they removed *any* of the nonconnected gates from the array, the device failed. Why, they have never been able to figure out. They guess that the gates either responded to "radio waves between components sitting right next to another, or they were somehow interacting through the power-supply wiring."[6]

But the largest puzzle is the "clock." Thompson "found the input circuit routed through a complex assortment of feedback loops. . . . [T]hese probably create modified and time-delayed versions of the signal that interfere with the original signal in a way that enables the circuit to discriminate between the two tones."[7]

"But, really," he concluded, "I don't have the faintest idea how it works."

Even though the capacity evolved by the FPGA is a basic one, that it was able to evolve it as its own, rapidly, in hardware, and in ways that exceed the grasp or design of its creator, is a truly eerie set of developments. Just what might such devices evolve if significantly greater resources were allocated to them? Let's see what happens to a Langton CA built of FPGAs.

Brain Builders: Back to the Future

In 1992, at age forty-four, following years of working in industry, a former physicist, Hugo de Garis, obtained his doctorate in computer science. The next year he was appointed to the Brain Builder Group of the Evolutionary Systems Department at the Advanced Telecommunications Research (ATR) Institute in Kyoto, Japan. He is now heading up a team at Star Lab in Brussels whose immediate aim "is to build artificial brains with billions of artificial neurons," to evolve a fully functional artificial brain within ten years, composed of one *trillion* neurons, and to "see brain-like computers become a trillion-dollar industry within twenty years."[8]

This was no pipe dream. With the Japanese government alone committed to providing $3 billion a year for ten years (starting in 1992), de Garis's Brussels group is part of an extraordinary multination effort to build a genuine artificial brain within the next few years. The principles of artificial brain building, using FPGAs in a massive, evolvable, cellular automaton, with gates set to function as cells, closely parallel the natural principles of embryogenesis that led to the human brain. For a two-dimensional prototype, genetic algorithms breed the update rules for the cells (the "chromosome" contains some 90,000 "genes") and there are 1 million cells. The entire structure evolves into a neural network, with about 90 cells per neuron. The result is a full-fledged, hardware-evolved neural net with nearly 40 million neurons in 32,700 linked clusters. Each module is evolved *in under one second*.

How do neurons arise from CAs? When Christopher Langton worked out his Q-shaped self-replicating CA (figure 8–1), he ended up with a structure with an uncanny resemblance to a biological axon surrounded by its protective myelin sheath. In fact, as shown in figure 8–7, as the CA updates and grows, the changing numbers inside the sheath form a propagating flow: *a signal*.

It took the de Garis team four years to design/evolve a selection of update rules for two-dimensional (and later three-dimensional) CAs that did not

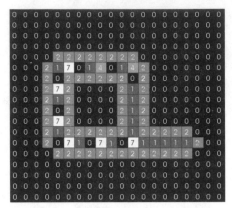

FIGURE 8–7   Langton self-reproducing cellular automaton from figure 8–1 with typical numerical state values. Sequences of 7,0 and 4,0 represent signals. These propagate around the Q toward the "tail."

merely reproduce to form a fungus-like blotch, like Langton's original reproducing CAs, but that would spontaneously generate genuine neural networks: linked "neurons," "axons" that transmit a single output value of one neuron to many others, multiple "dendrites" from other neurons carrying values that would be added together to form the inputs to a single neuron, nonlinear "transfer functions" in every neuron that would generate an output signal from the summed input and that made distinctions between excitatory and inhibitory neurons. It also needed to grow properly, and this required an ability to turn growing axons and dendrites to the left and to the right, to form branching ("fan-out") extrusions and to respond to collisions between paths— either to terminate one or both in a dead end or to form a synapse between them. (See figure 8–8.)

In an extremely clever adaptation of natural limitations, de Garis's group had the value of dendrite signals diminish with distance, while axonal signals remain unchanged. Since the length of any connection is an evolvable parameter (there is no requirement that two neurons be linked by the shortest possible

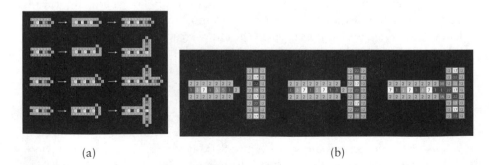

(a)   (b)

FIGURE 8–8   De Garis cellular automaton. (a) From top to bottom: extension (of a dendrite); left turn; left branch; T-branch. (b) Detail of a synapse forming.

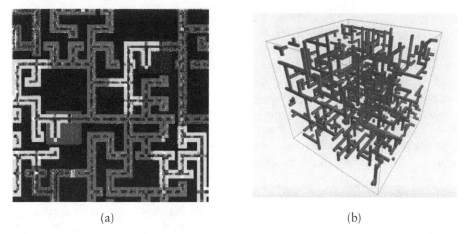

(a)   (b)

FIGURE 8-9   (a) Combinations of synapses and connections of various lengths forming part of a growing neural network. (b) De Garis cellular automaton forming a neural network module in three dimensions.

path), distance between the axon-dendrite synapse and the receiving cell serves as a proxy for connection strength. Eventually a network grows with combinations of these various elements, as shown in figure 8–9a.

In spite of its extraordinary complexity, the two-dimensional brain is seriously hampered by being limited to the plane, and de Garis abandoned it in favor of three-dimensional versions. The problem is not merely the vastly greater complexity that can evolve in three dimensions, but that in two dimensions, *growing axons and dendrites can't cross*—they must either form a synapse or terminate. Thus, the topology of three dimensions is qualitatively different from that of two. (Recent research suggests that three-dimensional processing is necessary to develop an ability to conceptualize and to visualize rotations in three dimensions.) The three-dimensional version resulted in structures that look like figure 8–9b.

Life: Set, Match, Championship?

Should De Garis's group, or any of a growing number of other groups around the world, succeed in evolving an artificial brain with 1 trillion neurons, it will be difficult to imagine that any human capacity will be beyond it. Groups already are generating commercial products to compute interactive, adaptive emotions for robotic dolls, and these require vastly fewer "neurons" and no evolution. If two HERs playing Hexapawn can fool us into thinking them human, will we be able to tell the difference between a trillion-neuron silicon-based brain and our own 20-billion neuron, carbon-based one?

Perhaps we will be able to tell the difference only because the silicon brain is *better* than we are at doing those things we consider the hallmark of a living mind. If so, it will surely seem that we have indeed discovered how the brain came into existence and how it works—and having succeeded, we will have gone on to do it one better. It will be hard to deny that we will have pene-trated into the heart of life itself, figured it out, and mimicked it.

Looking back at the territory we've covered, we will therefore arrive at the following conclusion: Man is a machine. The brain develops and organizes itself in a wholly mechanical fashion. To explain what we are and what we can do requires no appeal to anything beyond the natural potentials of dumb mat-ter interacting locally. No outside intervention, no ultimate HER superset, no intention, no design, no top-down programming is required. Everything that constitutes us and our existence is simply what mechanically had to be. There were no choice points in the processes that produced us; there are no options in the actions we take: The road not taken is nothing but a road impossible to take. There is nothing preventing us from generating—from forcing the evolu-tion of—alternate forms of "us," because there is nothing about us that sets us apart from mere mechanicality. We will have won the game of life at long last.

Or rather will we have lost?

9

IS MAN A MACHINE?

Everything we have presented so far would seem to imply "yes" in response to the question in the chapter title. One of the major objections to the reductionistic worldview has always been that life, human beings, the human fourth brain in particular, could not possibly have simply arisen by chance. Rather, its astonishingly clever interweaving of matter in so high a degree of order as to generate intelligence surely implies a designer. "No one could stumble upon a fine watch lying on a seashore," goes the argument, "and conclude that the watch assembled itself by the chance concatenations of happenstance." We have seen that devices cleverer by far than mere watches can and do assemble themselves in just that way.

"But who makes the rules?" you ask. The rules in the physical world are the physical laws that govern the local interactions of matter. Perhaps some "one" created them, in the way that de Garis hand-designed rules that would allow his creation to evolve toward the desired end of intelligence. But if so, the rule-maker long since took leave and left his creation to its own, sufficient, devices. For after the initial rules are in place, nothing we have discussed requires or even hints at the existence of processes that transcend a conventional understanding of dumb, mechanical interaction. "The cell is a machine," asserts the Nobel Prize–winning biologist Jacques Monod. ". . . man is a machine."[1] In showing that he is more particularly a *computing* machine, science has taken an enormous and exciting step forward, to be sure, but has thereby also made Monod's point even more convincingly.

But as exciting as such developments are to most scientists, and to most people with a scientific bent, they are also disconcerting. For where in a wholly deterministic universe composed of nothing but machines (however densely packed the one inside another like Chinese boxes) is there a serious place for freedom of choice? For free will? For meaning? For compassion? It's true we may well continue thinking that we are "free to choose" to be kind and good and loving (or not), but in such a universe such "freedom" is clearly an illusion.

We are not "free" at all to choose anything, including our beliefs, about this issue as about any other: We are simply deterministic cellular automata whose every successive state can be nothing but. (Even "random" fluctuations are just mechanical intrusions—they aren't really random, just extraneous.) If someone presents an argument to the contrary, you would be perfectly justified in saying "You only say so because you are utterly incapable of anything else." And he would be equally justified in responding "As are you." Whether you then proceed to have a beer and forget your differences or blow each other's brains out is likewise a foreordained sequence of events about which to think in terms of morality is completely absurd—if inevitable.

Because of this bleak portrait of human nature, when quantum mechanics—the major focus of Part Two of *The Quantum Brain*—first came upon the scene, there were many who intuited, without being able to show how, or at least hoped, that the true secret of life, as also of human nature and free will, was somehow bound up with the mysteries of the quantum. Quantum entities behaved like little beings possessed of free will, even if they were unimaginably tiny. In time quantum speculation became a veritable chaff-producing industry, and most serious scientists dismissed it. Most quantum speculation deserves dismissing because it consists of little more than analogies. These leap at one go from the scale of the unimaginably tiny to the scale of everyday life. But there is, in fact, another approach.

FROM SIMULATED TO QUANTUM ANNEALING

Whether in a spin glass, a cellular automaton, a neural network, or a genetic algorithm, optimization involves the settling of a multielement system into a "basin of attraction" of some "energy landscape" in accord with the principle, at every level from the molecular to the societal, that locally interacting elements spontaneously generate global order and systemic intelligence.

So, imagine two perfectly identical such systems, therefore with identical energy landscapes. Let us apply an identical annealing process (heat/cold regimen) to both, starting from the same initial state. We'll assume that the "randomness" (the heat, the extraneous jiggling) is also exactly the same for both. (For example, we can generate a table of pseudorandom numbers on our computer. We use this one sequence to determine the probability function for each neuron's output in both networks. Both networks therefore follow exactly the same trajectory of intermediate states in arriving at an identical final state.) We run our simulation once on both to confirm what we've described above. The results are as expected. Being identical, wholly mechanical systems, they both follow the same sequence of states.

Now we do it again, but we pull a fast one. Somehow (the details don't matter for the moment), we *"dig tunnels" in one of the otherwise identical landscapes*, connecting certain higher-energy "traps" to lower-energy basins. When we run both, are they still identical?

Clearly not. For from time to time, one of the two rolling balls can stumble upon a secret passageway to a lower-energy level. In the long run, in a complex enough landscape, this will result in two types of differences: The ball that cheated by opportunistically sneaking through tunnels might well achieve altogether better solutions (lower energy states) than the one that played by rules.

Even if both balls manage to find the same, absolutely minimal, energy state, by using tunnels, the first ball will most likely get there first.

Now let's add yet another twist, one about which you should think deeply. It may not make sense, but, again, for now, just take it at face value: Let's further suppose that while the annealing process (the convergence of our network/CA/spin system to an energy minimum) is entirely deterministic, the tunnels appear, disappear, then reappear elsewhere in an *almost entirely yet absolutely random* fashion. By this I mean: Each appearance of a tunnel, where it begins and ends, and the length of time it remains open is *absolutely and utterly random*: There is nothing whatever in the physical universe—neither at large scales outside the components that make up our experiment, nor at small scales within its inner workings—that causes these tunnels to appear and disappear. However, when a large enough number of tunnel appearances, durations, disappearances, and reappearances are analyzed, it is evident that, on average, they conform to some kind of pattern.

We can therefore paint a *probabilistic* portrait of how the tunnels enhance the annealing process, by the cheating they allow. From this point of view, the tunnels act on the system somewhat as heat does—introducing a kind of random "motion" that facilitates the settling of the system into a solution.

But we would be forced to conclude (because of our premise of absolute randomness to any individual tunnel's appearance, even if these appearances form typical patterns) *that our system as a whole is no longer deterministic.* The final states it settles into and how it gets there still conform, on average, to its fully deterministic twin, but the actual outcome of any one run is now no longer predictable. It will do things that are wholly unanticipated in the profound sense of having *no physical cause.* Furthermore, it will get results faster than could any deterministic version of itself.

Of course, all this is based on an absurd supposition. How can there be such things as magically appearing and disappearing tunnels that have no cause? Such a thing means that a physical system in one state could simply "snap" at will into an entirely different one, following no trajectory in the physical universe. It is precisely as ludicrous a notion as the idea of cars trapped in a dead-end valley in the Italian Alps suddenly disappearing and reappearing somewhere in Switzerland. Note, too, that in this analogy, the term "tunneling" is not quite apt. To go through a tunnel, one must *travel* through it, point by point. What we are speaking of here is rather more like teleportation: Intervening points are never occupied.

But what I have described here *is* part of quantum mechanics. If the network of which we are speaking is in fact a network of electron spins in a lattice of atoms, for example, then there *do* exist such tunnels between otherwise mutually inaccessible parts of an energy landscape. (See figure 9.1.) The precise meaning of this—in all its truly unbelievable weirdness—is the subject of the next section of the book. But for now, let's take a quick peek ahead because of its immediate application to what we've discussed so far.

In April of 1999, two papers on quantum annealing appeared. One used simulation techniques to quantify what I've just described in rough terms: that if teleporting between states is allowed in a system, its computational function is strikingly enhanced. The researchers, Wolfgang Wenzel and Kay Hamacher at the Institute for Physics at the University of Dortmund in Germany, found that, in a traveling-salesman-type problem, quantum annealing found a 35 percent better solution and did so ten times faster than classical annealing. Applied to a model of an actual spin glass composed of electrons, quantum annealing was about 100 percent more efficient. Finally, they applied the same technique to a chain of spins, "one of the hardest discrete minimization problems known." The most sophisticated and complex nonquantum techniques were unable to arrive at solutions with less than 20 percent error no matter how long they were run. Quantum annealing quickly achieved less than 1 percent error.[2]

The second paper, "Quantum Annealing of a Disordered Magnet," examined ~10^{23} spins in a rare-earth alloy spin glass. The authors explain that by contrast to traditional simulated annealing, "[q]uantum tunneling provides a different mechanism for moving between states, . . . quantum annealing hastens convergence to the optimum state."[3]

The quantum protocol caused relaxation of the spins into a steady state from three to thirty times faster than when no teleporting of states occurred. (Teleporting was induced by applying a magnetic field.) Furthermore, the researchers estimated that the final state was almost certainly the absolute energy minimum, or extremely close to it. But more important, the state of the magnetic material was altered—it retained a memory, so to speak, of its "impossible" history.

In other words, the presence of ultramicroscopic quantum events occurring locally among some 10^{23} individual elements generated a persistent *global* difference in their overall organization. Neither was this difference trivial or fleeting: It affected a major characteristic of the material at everyday scales, and this difference, achieved via seemingly impossible means, persisted for time scales vastly longer than the quantum events themselves.

So let's speculate a bit, and look ahead: What if the evolving FPGAs in Hugo de Garis's artificial brain could be built on chips that would seek energy minima not merely via the already amazing but nonetheless wholly mechanical

FIGURE 9–1   Direct access of low energy state via quantum mechanical tunneling.

processes of optimization we've discussed, but via their quantum analogs? There would surely be a dramatic increase in the rate at which they evolve as well as in their sheer problem-solving power.

But would not such arrays, as they evolved, be likely to take unexpected advantage of the unique quantum states available to them—just as Adrian Thompson's FPGA found ways (which he still can't figure out) to make use of gates that were never even wired up? Would they not also be somehow "different"?

We might also wonder: Would such plausible results yield computers that exceed not merely quantitatively but, in their very essence, the capacities of the human brain? If, as the vast majority of neuroscientists today insist, the human brain is merely an extraordinarily complex, but still wholly deterministic machine, then computer brains that incorporated the quantum advantage would indeed have a fundamental—one might almost say magical—capacity that we humans lack.

Or perhaps not. It might rather be the case that what we are getting a glimmer of now in artificial systems will show us the way to discover much the same within ourselves—just as the development of neural networks shed the first clear light on how networks of neurons work. It would be fine poetic justice, should our construction of thinking machines that turn out to be not really machines at all enable us, for the first time in four hundred years, to see from a scientific perspective that neither are we. To see how this may indeed be so, we turn to the mysteries of the quantum world and their unexpected intersections with the human brain.

PART TWO

―――――――

MIRACLES

10

THE INSOLUBLE SOLUTION

*Miracles are hard to accept. Abandoning the causal conception
. . . threatened our basic ideas about nature . . . and pushed pure
physics toward the realm of metaphysics. But most younger
physicists accepted this bargain and struggled to follow quantum
mechanics as it grew.*
—Eugene P. Wigner, Winner of Nobel Prize, Physics, 1963

BRUSSELS, OCTOBER 1930

"A struggle of Titans over the last riddles of the universe" was how one writer put it. For years, Albert Einstein had fiercely battled Niels Bohr's emerging take on quantum mechanics. Bohr had at first triumphed, but Einstein fought back and at long last seemed to have won. But this evening, in a flash of genius, Bohr appeared to have turned the tables on him.

It was impossible. Only a child could truly love the claims of quantum mechanics that Bohr championed: that physical particles lacked fixed positions and velocities (more precisely, momentum, the product of an object's mass and velocity)—that to the extent you knew one, the other was vague—not merely that one's knowledge of it was vague; that entire atoms could have a place in both space and time as solid objects and yet infinitely spread out as ripples in the fabric of some undetectable medium; that the universe was filled with events that had no cause whatsoever. Yet these were in truth its major claims, and to date no experiment had ever produced results to contravene it; and there were other experiments, exceedingly important ones, whose results could be explained in no other way. The mystery of radioactive decay, of the photoelectric effect, of why certain hot bodies appeared black—these last three loose ends from classical physics all yielded to quantum theory alone.

This was not a simple battle over some abstract philosophical or mathematical formulation, as it must surely seem to outsiders, nor a matter of mere technology. Einstein was explicit: Upon its outcome rested the entire enterprise of science, and its triumph over superstition, religion, and magic. If Niels Bohr was right, then reality itself would make no sense: Einstein saw that in the end, quantum theory required that the entire universe, no matter how remote its parts from each other, be woven together into a seamless unity, somehow orchestrated in concert; that events everywhere were being coordinated outside the bounds of causality by some unimaginable "influence": God playing dice. Seeing into this dark secret at the heart of the quantum, Einstein was certain beyond any doubt that the theory had to be fundamentally flawed, however seductively accurate it appeared in tests to date. It was only a matter of time before the flaw was uncovered.

Among the physics elite, a pure meritocracy prevailed. The meetings at Solvay, sponsored every few years by an extraordinarily wealthy Belgian industrialist and amateur physicist, had become the physicists' de facto tourney royal—jousts and all. Two years before, in 1928, at the previous Solvay meeting, Bohr's arguments had been simply brilliant, and Einstein, who rarely took second place to anyone, had been soundly defeated. Paul Ehrenfest, a friend of both men and a major contributor to physics in his own right, would later describe the climax of that earlier conference:

> Brussels—Solvay was fine! . . . Bohr . . . [a]t first not understood at all . . . then . . . defeating everybody. . . . Like a game of chess, Einstein all the time with new examples. . . . Bohr from out of philosophical smoke clouds constantly searching for the tools to crush one example after the other. Einstein like a jack-in-the-box; jumping out fresh every morning. Oh, that was priceless.
>
> But I am almost without reservation pro Bohr and contra Einstein. His attitude toward Bohr is now exactly like the attitude of the [opponents of relativity] towards him. . . .[1]

Einstein had left the field of battle in retreat but certain that, even so, he was correct and quantum mechanics in error. He afterward wrote to Erwin Schrödinger—one of the very greatest of the developers of early quantum theory: "The soothing philosophy—or religion?—of Heisenberg-Bohr is so cleverly concocted that for the present it offers the believers a soft resting pillow from which they are not easily chased away. Let us therefore let them rest. . . . This religion does damned little for me."[2]

However great his distaste for quantum mechanics itself, Einstein tried to remain personally dispassionate. That very year he nominated both Schrödinger and Heisenberg—Bohr's most important ally—for the Nobel Prize. But he added that "it is still problematical how much of the grandiosely conceived theories of the two last-named scientists will survive."[3]

To this sixth Solvay conference Einstein had come prepared with a definitive refutation of the quantum enterprise. And two nights earlier, the world's greatest minds gathered around him and Bohr, he had triumphed.

Einstein had spent the previous year developing a "thought experiment" of exceeding subtlety, the mere devising of which seemed to reveal a necessary contradiction at the heart of the quantum theory. It was a move of dazzling beauty and simplicity—indeed, in a class never before dreamed of—such that the outcome of the rest of the game, and the match, seemed decided at a stroke. In a sentence: Einstein showed that the notorious quantum "uncertainty" between time and energy—a necessary counterpart of uncertainty between position and momentum—could be avoided by weighing an object in a box that has lost a photon of light when a timer briefly opens a shutter. The timer tells you *when* the photon was let go; weighing the box tells you *how much energy* the photon has—because of the famous relationship between energy and mass: $E=mc^2$, the pinnacle of Einstein's *Special* Theory of Relativity. Bohr, a genial, unflappable Dane, paled visibly when Einstein first laid out his masterstroke. Wrote spectator Léon Rosenfeld:

> To Bohr, this was a heavy blow. He . . . walked from one person to another, trying to persuade them all that this could not be true, because if Einstein was right this would mean the end of physics. But he could think of no refutation.
>
> I will never forget the sight of the two opponents leaving the university club. Einstein, a majestic figure, walking calmly with a faint ironical smile, and Bohr trotting along by his side, extremely upset.[4]

The great Danish physicist spent the entire night awake, searching for a crack in Einstein's diamantine logic.

Shortly before dawn he found it. Later that morning Bohr showed Einstein his discovery. Rather amazingly, Einstein had neglected to take into account the peculiar effects of gravity required by his own theory of *General* Relativity, a theory of even greater scope than Special Relativity. General Relativity showed that time itself slows down in a gravitational field. When the photon leaves the box, the mass of the box-plus-timer-plus-photon gets smaller, the nearby gravitational field goes down, so the passing of time in the vicinity of the box speeds up. Using the equations of General Relativity, Bohr showed that the change in mass caused by the loss of a single photon is reciprocated by a change in the scale of time such that the product of the uncertainty of the time and the uncertainty in the energy exactly equals that minimum uncertainty predicted by quantum theory. If you insisted that your time measurement was absolutely perfect, then your weight (energy) measurement would be absolutely uncertain—a range of plus or minus infinity!—and vice versa.

It was a terrible irony. If Einstein was right about quantum mechanics, he was utterly wrong about a theory the completion of which he himself called "the greatest satisfaction of my life," that which had led his colleagues to hail him as "the new Copernicus." The shock was unimaginable. The succeeding morning would not bring yet another reversal of fortune. Nor any mornings at Solvay thereafter.

As always, Einstein was gracious in defeat. He again wrote the Nobel Prize committee, describing Schrödinger and Heisenberg as "the two men who above everyone else deserve the Nobel Prize for physics." And his reservation was far more restrained: "I am convinced that this theory undoubtedly contains a piece of definitive truth."[5] But, gracious though he may have been in withdrawing from the field at Solvay, he had hardly surrendered.

Five years later, in 1935, Einstein and two colleagues, Boris Podolsky and Nathan Rosen, published a brief but overwhelmingly brilliant piece—referred to ever since as "EPR," for their initials—asserting that quantum mechanics was true, as far as it went, but *incomplete:* It was but a "piece" of truth. The portrait of a world in which causality no longer held absolute sway was not a true principle but must be an artifact of our ignorance of elusive, exceedingly subtle causes. The absolute truth of causality—that *nothing* ever occurs without wholly sufficient material causes that can yield but one result—is a vision of reality upon which all of science was based, and once it came to Einstein—when he was just a little boy, sick in bed—it never left him: "I experienced a wonder . . . as a child of 4 or 5, when my father showed me a compass. The fact that the needle behaved in such a definite manner . . . I remember to this day . . . the deep and lasting impression this experience made on me. There had to be something behind the objects, something that was hidden."[6]

The memory not only expresses a worldview he never abandoned, it foreshadowed his greatest contribution: the theory of relativity. For although it is not widely known to those unschooled in physics, the Special Theory of Relativity is primarily an astoundingly deep explanation of magnetism. What is electric to a moving observer appears as magnetic to one at rest. From this, Einstein showed, must follow everything we know as "relativity"—shrinking objects, time dilations, atomic bombs.

The EPR paper is one of the most elegantly reasoned in all of intellectual history.[7] It showed that if quantum mechanics *was* complete—meaning that its effects were produced not by undetected causes but by no causes whatsoever—then the very physical condition (the "reality," as Einstein termed it) of, say, two or more separate events (or things) was contingent upon our measuring and knowing at least one of them; more absurdly yet, under certain circumstances that Einstein and his colleagues teased out, their physical reality could be shown to be contingent upon our mere *intention* to measure one! And this would obtain even were the second object so distant from the yet-to-be-measured first that measuring the first object *here* could not possibly have an effect, soon enough, *there*. The second item could be in the Andromeda galaxy, and this impossible, instantaneous "influence" would still hold. The influence could even run backward in time. No one other than mystics and madmen could entertain such patently absurd notions.

The mathematics behind Einstein's argument was flawless. Either quantum theory was fundamentally incomplete, or reality was weird and contradictory beyond imagining—if backward influence could exist, its features would even depend on what one's intentions toward it were. In Einstein's dry formulation, "No reasonable definition of reality could be expected to permit this."[8]

Einstein sent Bohr a copy of the paper prior to publication. Bohr was devastated. His colleague Leon Rosenfeld described how they reacted:

This onslaught came down upon us as a bolt from the blue. . . . [E]v-erything else was abandoned: We had to clear up such a misunderstand-ing at once. We should reply by taking up the same example and showing the right way to speak about it. In great excitement, Bohr immediately started dictating to me the outline of such a reply. Very soon, however, he became hesitant. "No, this won't do, we must try all over again . . . we must make it quite clear . . ."

So it went on for a while, with growing wonder at the subtlety of the argument. . . . Clearly we were farther from the mark than we first thought. Eventually, he broke off with the familiar remark that he "must sleep on it."[9]

But empirical tests of the EPR argument were out of technological reach. As a result: "For decades, paper after paper was written in an attempt to resolve . . . [the EPR paradox], but not for thirty years was any real headway made on the matter.[10]

Extraordinary claims require extraordinary evidence, and the claims of quantum theory were extraordinary beyond imagining. In the absence of defi-nite evidence one way or the other, Einstein's argument against quantum weirdness was convincing to many. To others it was not. Perhaps reality *is* truly weird.

THE NEW PRIMORDIAL SUBSTANCE: NOT MATTER BUT *INFORMATION*

In the meantime, the other sciences were undergoing rapid developments of their own, unrelated to the startling events in physics. In part this was simply because the mathematical foundation of the new physics was so complex and esoteric that few apart from physicists and mathematicians really understood them. But most sciences had no need to relate to the subatomic world any-way—what does electron "uncertainty" have to do with, say, sociology? The other sciences were experiencing one triumph after another not by *abandoning* a long-held belief in mechanical determinism but by for the first time *embrac-ing* it. Ironically, scientists in other fields saw themselves thereby as at last emu-lating physics—the gold standard of scientific inquiry.

Jacques Monod spoke unflinchingly on behalf of most scientists in his claim that "Man is a machine." But Monod either didn't know or (more likely) dismissed as irrelevant to biological systems the fact that physics was moving on into utterly confounding territory. Physics itself was now saying that, far from being reducible to "simple mechanical interactions," *matter itself*—the very foundation of the machine viewpoint—was looking less like a machine and "more like a thought."

The notion was tough to swallow, but "Get used to it . . ." became the working physicist's motto, ". . . and let the philosophers scratch their heads." Even if the tangible "stuff" supposed to make up the natural world was fading away into unreality—into sheer "information"—tangible progress was rapid. So what if, say, transistors, made no sense; they worked pretty damned well.

In response to quantum theory, three distinct camps emerged within the sciences. First among peers were the originators of quantum mechanics, along with others capable of carrying quantum theory forward—or of disputing it— on a rigorous mathematical basis. In private they discussed its philosophical implications, but only on rare occasion did they allow their musings about "what it all meant" to be released in print. A few were quietly mystical, a few wondered whether in quantum mechanics lay the secret of life, of will, of consciousness, even of God. Most, however, believed that in the end, mechanical determinism was bound to win out, somehow. All agreed that the question would be settled not by philosophy but by mathematics and experiment.

The second camp consisted largely of a second tier of scientists, philosophers, and poets, who saw in quantum theory a long-hoped-for opening to a world of mysticism and did everything they could to widen it—an opening that Einstein likewise saw and wanted to prove illusory. Allying with a few first-tier physicists—Heisenberg, Wolfgang Pauli (a close friend of the mystical psychiatrist C. G. Jung)—these scientists began a process of open philosophical speculation that irritated most serious physicists. The rankest of speculators blended physics and (usually) Eastern mysticism without reservation, but were rarely able to express their ideas in mathematical form; nor could they conceive of tests that could prove them wrong. But they intuited and guessed and analogized at length, sometimes wildly, sometimes presciently, in an attempt to draw out the large-scale antimechanical implications of quantum mechanics.

Often these implications had a distinctly trendy bias. Author Dana Zohar, for example, has gone so far as to find in quantum mechanics the basis for a "new morality" in which adultery is no less a virtue than faithfulness. "An evolving 'quantum morality' . . . is necessarily . . . pluralistic," she writes. "Our free and undetermined moral choices . . . are like an electron's virtual transitions."[11]

The third and largest camp is the vast body of competent scientists of every stripe without training in quantum mechanics, and often without interest in it. Theirs is a commonsense point of view that works quite well. "Whatever 'mysteries' may or may not lie within the domain of quantum physics," they say, "these do not concern us. Once you begin studying phenomena that require trillions of subatomic events, as we do in studying chemistry, biology, medicine, all the mystery averages out and you are left with pure mechanism. This is a fact that has never been controverted. The world that we study—including our own everyday lives—would behave not the slightest bit differently, nor less mechanically deterministic, depending upon whether quantum theory is right or wrong, or upon what it 'means.' For that reason we continue to treat everything as a machine—cells and man and his brain included, because when we do so, we make enormous strides in our understanding and our ability to predict, and we don't we trip up. *All* of modern medicine, for example, depends upon the machine point of view that we adopt; almost *nothing* in modern medicine—hardly a single advance—has ever required that we abandon it. Quantum theory is simply irrelevant to anyone but physicists, theoretical chemists and New Age philosophers—and only the first two understand it—sort of." In short, "the reductionist hypothesis may be still be a topic for controversy among philosophers, but among the great majority of active scientists I think it is accepted without question."[12]

For example, the quantum "uncertainty" in the position of a typical atom that's part of a typical cell might be on the order of about 10^{-8} centimeters (two one-hundred-millionths of an inch). The smallest organelle within that cell might be 10^{-5} centimeters across (two one-hundred-thousandths of an inch). The quantum uncertainty of that atom, in other words (and that of any other constituent atoms), would amount to at most one-thousandth of the size of the organelle. Think of it, if you like, as a "vibration," like the agitation caused by heat, only smaller. It's hard to imagine how such a slight additional vibration could have any significant effect in living processes, especially when everything is vibrating in uncoordinated fashion.

Quantum uncertainty is like the random motion of air molecules in a room. Because the motion is "random," there is a chance that all the molecules will just happen to dart into the extreme upper left corner, all at the same time, suffocating us. But it is a statistical certainty that such a thing is not going to happen. Quantum uncertainty—if there really is such a thing—may allow for "freedom" philosophically, but to depend on it as an explanation for freedom at the scale of our everyday experience is actually to make the case for determinism: By the time all the untold trillions of individual "freedoms" are added up and averaged, the proportion of freedom left to something large—a human being, a human brain—is so tiny a fraction that it would be utterly beyond detection. Whether such statistically enforced determinism is really significantly different from mechanical determinism—and whether you ought to become a mystic because of it—is a debate we can safely leave to the philosophers and poets.

But in the last ten years, a renewed debate over reductionism and determinism shows clear signs of breaking out into the open. A cascade of advances has emerged from physics, neuroscience, and computer studies that show the brain's actual operation to be very different from it has long been thought.

First, as we've discussed in Part One, the human brain is proving to look like a self-organizing computer that teaches itself and learns by experience: once thought impossible. In many ways, these discoveries only reinforce the idea that everything the brain does can be explained entirely via mechanism. But it turns out that self-organizing ensembles of every kind have in common a tendency toward *chaos*. (More precisely, such systems demonstrate so-called nonlinear dynamics, of which chaos is a particular form.) Strictly (i.e., mathematically) speaking, chaos does *not* mean disorder; the proper term is "deterministic chaos." Chaos in this sense means a sequence (say of numbers) that is strictly mechanically determined by some fixed rule, but if you don't know the rule, you cannot figure it out just by looking at the numbers. Furthermore, no two numbers in a chaotic sequence are *ever* identical, though the sequence returns at roughly orderly intervals to nearby numbers. For these reasons, a chaotic sequence looks random, but it isn't—it is 100 percent mechanical.

Popularizations of chaos theory often claim that a chaotic sequence is in some sense "free," and that the presence of chaos in human physiology proves the existence of free will. This is incorrect. Chaos merely sets a limit on our capacity to deduce the rules—hence the mechanical causes—in retrospect. It does not in any way make those rules disappear. Indeed, chaos arises wholly because of them.

Chaotic behavior can arise only in systems whose state at any moment in time, for instance, is some function of its prior state, though not in every such system. Such systems are "iterative" in the way that each error-correction cycle of a back-propagation network is, or each update cycle is in a cellular automaton. The best-known example of an iterative system is the "compounded return" on your bank account. Though the interest rate is constant, the amount of additional money you earn each cycle changes as a function of how much money you have already accumulated. The growth of your account—presuming you let it be—is *nonlinear,* not straight line because the base upon which the interest is calculated changes as a consequence of the interest added to it during the prior update. The growth is nonlinear, but it's also "monotonic"—it goes only up, it never declines or cycles or jumps about. This kind of non-linearity is not and cannot be chaotic.

But there are many iterative, nonlinear systems that do illustrate chaos. Here is a simple example: Consider the sequence of numbers generated by the equation $y = 3(x - x^2)$, as we allow x to take on different values, one after the other, starting with 0: 3, 0, –9, –24 . . . and so on. These values (as we graphed them in algebra class) all lie on a parabola. If we included every value of x between 0 and 1, and between 1 and 2, and . . . the values of y would form a smooth curve. But now change the "3" to "3.6," and make the equation iterative—in other words, at each step use as x the value of y generated by the *prior* value of x. "Seed" the equation with an initial value of 0.5. Now the sequence looks like 0.9, 0.324, 0.788486, . . . The values of y will always remain between 0 and 1, no two values will ever be the same, and if you look at any snippet of the sequence, you will never be able to deduce the rule that generated it. The sequence is deterministic yet chaotic.

There is another feature of chaos that will prove extremely important in the following discussion of quantum mechanics and brain function: namely, chaotic systems amplify extremely small differences in initial conditions into dramatically divergent outcomes. For example, if instead of seeding our iterative equation with 0.5, we seed it with 0.500000000000000000001, then within a handful of iterations, the sequence will be completely different than were we to seed it with 0.50000000000000000000. As we will see, this allows for the possibility that very tiny effects, rather than being averaged away, can be, on the contrary, preserved and enlarged—but only in systems that are in every respect and at every scale *iterative*. Self-organizing systems are just that—precisely because iteration allows for the internal feedback that causes such a system to spontaneously establish order global order. As we will also see, systems of coupled, iterative *oscillators,* in particular, are even more sensitive to initial conditions, and these are just that of which the fourth brain is composed.

Second, "experiments have now shown that what bothered Einstein is not a debatable point but the observed behavior of the world."[13] All of the material universe does in fact seem knit together into a single, mutually influencing whole—or, perhaps, a whole everywhere influenced simultaneously, and coordinated, by "something" not itself a physical part of that universe. As mentioned before, it seems that at the foundation of matter lies not mere "stuff" but pure "information." Such a statement should seem senseless—what can it mean to say that "matter" is "information"? For now, just suspend judgment; to understand what this means clearly we will first have to assemble certain building blocks of understanding.

Third, quantum effects can now be produced on very large scales—well into the visible—raising the speculation that perhaps the brain *could* embody quantum events large enough to have a significant effect on day-to-day functioning—a speculation most qualified physicists reject for reasons we will discuss—for the most part, good reasons, but not entirely.

Finally, advances in neuroscience allow us to penetrate so far down into the microscopic substructure of the brain—well below the level of neurons themselves, into *their* structure—that not only might we find quantum effects there, physicists fully expect them to be present. The question is whether it's possible that these effects aren't just averaged away.

These four advances suggest the following question: Even if quantum effects are present within brain cells at a sufficiently large scale to affect neuronal substructures, is there any reason to suppose that these effects possibly influence everyday life?

One answer is: There exist structures in the human *brain that appear perfectly designed to capture quantum effect and amplify them via chaos,* because in order to self-organize, they have adopted an iterative form at every level—chemical, intracellular, network, and even social. If so, the actions generated by the brain, and of human society as a whole, would share at least some of the absolute freedom, mysteriousness, and nonmechanicality of the quantum world. Can this claim really be right? To find a plausible answer, we need to begin where Einstein went wrong.

11

EPR:

WANTED: DEAD AND ALIVE

We know that the deterministic world of classical mechanics
does not exist. Once we have bitten the quantum apple, our loss
of innocence is permanent.
—R. Shankar, *Principles of Quantum Mechanics* (1994)

In 1962 Richard Feynman presented the final third of his famous series of lectures on physics at Cal Tech. The series served nominally as the two-year introductory physics sequence for entering undergraduates. Because the number of attendees stayed roughly the same, Feynman didn't notice that by the second year the audience was composed primarily of upperclassmen, graduate students, and faculty able to appreciate the enormous beauty, power, scope, and depth of his presentation. The lecture series has become an icon of modern physics—incomparably elegant, incomparably casual, the quintessence of Feynman's style.

This last third was devoted exclusively to quantum mechanics. Feynman opened his exposition with a famous thought experiment, because, "[i]n reality, it contains the only mystery . . . In telling you how it works, we will have told you about the basic peculiarities of all quantum mechanics." However, the phenomenon in question "is impossible, absolutely impossible, to explain in any classical way," he cautioned. "We cannot make the mystery go away by 'explaining' how it works. We will just tell you how it works."[1] Here's what he was talking about.

COUNTING PELLETS

Imagine a gun that sprays pellets through a barrier with either one or two vertical slits. (See figure 11–1.)

Figure 11–2 shows the single- and double-slit arrangements viewed from the top down. To the right of each setup is a graph of how many pellets hit.

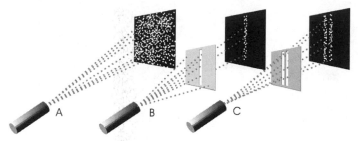

FIGURE 11–1   Pellets striking a target. (A) No barrier; (B) Single-slit barrier; (C) Double-slit barrier.

Naturally, if we close off one of the two slits, we get the same kind of result as if there were only one slit to begin with. The center of the single peak is, of course, lined up with the center of the single open slit, as shown in figure 11–3.

Now we do a similar experiment with water in a rectangular tank. (See figure 11–4.) A device at the left end of the tank taps out parallel waves that travel toward a detector at the right end. This detector measures the maximum wave height at each point along the right-hand side of the tank. A graph farther to the right shows the height distribution. We first try it with no barrier (right-hand figure) and with a single "slit" barrier—in this setting, the equivalent of a slit is a slot—inserted into the tank between the wave generator and the detector.

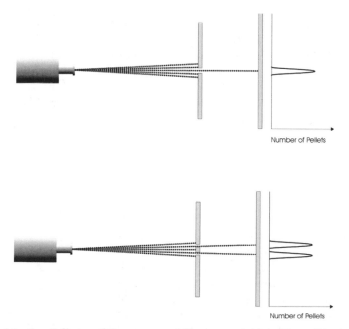

FIGURE 11–2   Pellets striking target: midline cross-sectional (top-down) view with graph of density.

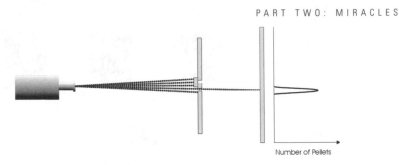

Number of Pellets

FIGURE 11–3   One slit closed.

The wave intensity (maximum excursion of the water, up or down) falls
off along the detector, because the energy of the wave disperses as it travels—
like heat from a radiator. The curve is broader than the single-slit distribution
curve for pellets but similar in shape. Figure 11–5 shows what happens when we
insert a two-slot barrier. The results (early shown on top and later on bottom) as
the waves propagate differ dramatically from our experiment with pellets.

Waves consist of regularly alternating peaks and troughs, above and below an undisturbed surface. As waves intersect, they combine. At the detector,
we therefore find bands of double-height peaks alternating with double-depth
troughs separated by areas where peaks and troughs cancel out—in the ideal,
extending ever more faintly forever: an "interference pattern."

With a double slit (slot), the pellet results and wave results are utterly distinct. It is inconceivable that a bunch of pellets shot through two slits could
produce an infinitely wide (if rapidly fading) interference pattern; and it is
equally inconceivable that two circular, continuous, infinitely radiating, inter-
secting wave trains could produce two sharply discrete nearby peaks.

This is because "waves" and "pellets" are not just different "things,"
they are completely different categories. A wave is a moving pattern assumed
by a collection of things; a pellet is a thing proper. That the waves and pellets
both seem to be "moving" rightward is likewise a misleading statement. Only
the pellets are really moving rightward. The water molecules (the true things)
of which the wave is made do not move rightward at all. They only move ver-

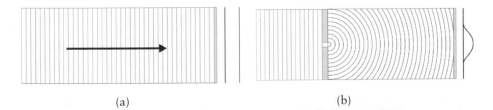

(a)   (b)

FIGURE 11–4   Top-down view of two wave tanks with detectors at the far (right-
end) wall of each that measures the maximum wave height at each point along the far
wall. To the right of each tank is a black line that indicates the relative height of the
wave along the detector. Uniform waves are generated at the left-end wall of each
tank. (a) No barrier. Parallel waves reach the detector with the same height at every
point. (b) Barrier with a single slot. Circular waves form from what has now become a
"point" source (the slot) and so are weaker where they must travel farther.

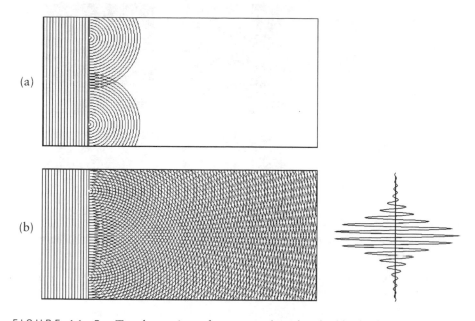

FIGURE 11-5   Top-down view of a wave tank with a double-slot barrier and a detector at the far end to measure the wave intensity along the far wall. (a) Shortly after the parallel waves have hit the barrier. The transmitted wave consists now of two "point" sources generating identical waves in phase. When they meet, they interfere. (b) When the waves reach the detector, alternating regions of constructive and destructive interference form the oscillatory intensity pattern shown to the far right. The dying out of the interference pattern is exaggerated.

tically—that is, they oscillate in place. What moves forward is the pattern. Just as waves are not "things," but "patterns," they don't really "move" at all; we therefore say instead that they "propagate."

Also, while pellets have a single discrete location, waves are spread out in (eventually) huge circles—infinite circles, in the ideal case. They may register more strongly at certain points than others, but they have no single location— they're everywhere, so to speak. A ball ricocheting around in a closed box has a different but precise and single location in the box at all times. A wave ricocheting around in a closed box fills the entire box at all times with a changing pattern caused by its interference with itself.

Finally, a pellet is made of actual stuff, whereas a wave proper is "made" of nothing, since it's just a pattern. It's not really even "made of water." Rather the medium in which it propagates is water. (It sounds like we're picking nits, but the distinction will help when we try to understand the relationship between matter—or energy—and information.) If you follow one wave and think of it as "made of water" and "moving," then you'd have to say that the water of which it is made is constantly exchanged as it moves. The horizontally moving pattern oscillates some water molecules vertically in place, leaves them behind and moves on to the next molecules.

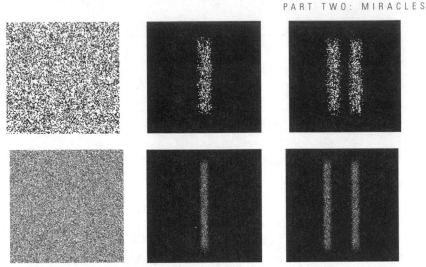

FIGURE 11-6   Top: Sand-sized particles striking a target. (left) No barrier. (middle) Single-slit collimator. (right) Double-slit collimator. Bottom: Electrons striking a target, expected results. (left) No barrier. (middle) Single-slit collimator. (right) Double-slit collimator.

In short, if something generates interference patterns, it must be a wave: a substanceless pattern propagating in some medium capable of oscillating or vibrating in place.

Now let's repeat the experiment with a sandblaster, so the particles we're firing are much smaller. What do we see now? The results are as expected, from (a) no slit, (b) one slit, and (c) two slits, shown in figure 11-6 (top).

Finally, let's use really small particles this time: electrons. Figure 11-6 (bottom) shows what we expect to see.

But figure 11-7 shows that what actually happens is an interference pattern! So, electrons cannot be discrete, localized objects at all, they must merely be spread-out propagating patterns in some medium. Let's examine these results more carefully.

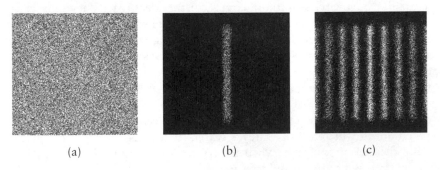

(a)   (b)   (c)

FIGURE 11-7   Electrons striking a target: The actual results are (a) No barrier; (b) Single-slit collimator; (c) Double-slit collimator.

We notice three very weird things right away. First, these "waves" are actually made up of individual particle strikes. Second, each individual particle strike seems located at random, but in the aggregate, the strikes conform to a wavelike distribution. When we measure the density of strikes as a function of distance from the midline, we discover that this density is precisely equal to the *square* of the height of a genuine wave interference pattern. This makes even more baffling the question of just what are the underlying, interfering "waves." They certainly can't be the "square root of a bunch of alternating plus and minus electron amplitudes," since that's just a jumble of words meaning nothing. But that's what the math says they are. Third, pick a location along the detector near the midline where, with two slits open, there are no (or very few) strikes. Now examine that same location with one slit open. You can find places where with one slit open there are strikes. In other words, at these locations opening an *additional* source of electrons *decreases* the number of strikes!

With all these weird things going on, it seems as if somehow, the electrons are "interfering" with one another and sometimes "canceling each other out"—whatever that means. Let's try to settle this. We calibrate our electron gun so that there is never more than one electron between the emitter and the target. We watch them arrive one at a time. Let's guess what ought to happen.

First, we know that when a single slit is open (see figure 11–6d above), electrons behave exactly as they ought—like pellets. That's why we see the "shadow" of the slit on the detector. So, simpleminded as it may seem to say it, let's do so: "A single electron that strikes the detector must travel through the single open slit." Sounds reasonable—especially since with no slits, no electrons get through.

Second, opening another slit can have no effect on the *one* electron that goes through the first slit. Furthermore, a single electron traveling through any one slit, whichever it is (and obviously, it can travel only through one, either the first or the second), can hit the detector only somewhere within a certain distance from the midline. Even if it scatters a bit off the edges of the slit (that's what causes the fuzziness of the "shadow"), there's a limit. It can't make sharp angular turns and show up a mile to the right. And since it's the only electron around, it has no others to bounce off of and create weird effects.

So, a single electron sent toward two slits and detected will have passed through one slit or the other, and will strike the screen in one of two narrow parallel vertical regions demarcated by the faint dotted lines in figure 11–8.

As there is no possibility for two successive electrons to interfere with each other, we must find that as the number of individual strikes grows, they will form the pattern shown in figure 11–9.

But "must," like Einstein's "reasonable definition of reality," scarcely means what we think it "should"—in the quantum world. Exactly such an experiment, with electrons, was first carried out in 1989 by a team of Japanese researchers at Hitachi, Ltd. The actual results are shown in the bottom row of figure 11–10.

FIGURE 11-8   A single electron is expected to strike within the dotted lines determined by all the possible trajectories through either one of the two open slits.

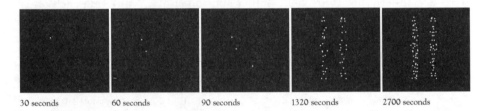

30 seconds   60 seconds   90 seconds   1320 seconds   2700 seconds

FIGURE 11-9   Many single electrons are expected to strike within the dotted lines
determined by all the possible trajectories each individual electron might take through
either one of the two open slits. (The times are hypothetical.)

Now, if there is only one electron at a time traveling from the tip of the
emitter to the target, how in the world can electrons interfere with one another? With what is the first electron interfering? With electrons that have yet
to be emitted? Eerily, this experiment will work no matter how much time
there is between the individual electrons—even years.

"OK," you say, "maybe electrons really aren't particles at all, even if
when they strike the screen they leave a discrete mark. Maybe they're waves,
somehow, pure and simple. After all, no one has ever actually seen an electron.
I've heard people talk about how the solar system model of the atom is wrong,

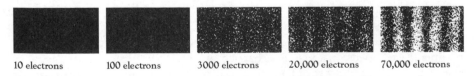

10 electrons   100 electrons   3000 electrons   20,000 electrons   70,000 electrons

FIGURE 11-10   Actual results showing build-up of electron interference over
time. The pattern builds up at the rate of about 1,000 electrons per second, more than
adequately slow to ensure that only one electron at a time traverses the space from the
gun tip to the detecting screen.

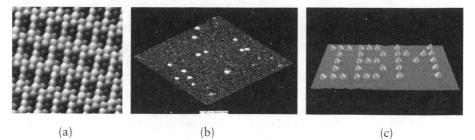

(a)   (b)   (c)

FIGURE 11–11   A "scanning tunneling microscope" produced these actual pictures of various atoms. (a) Oxygen atoms (dark) on the surface of a lattice of plati-
num atoms (pale). (b) Gold atoms (white) scattered at random on a mica surface.
(c) "IBM" spelled out by placing individual atoms in fixed locations.

that the electron is more like a cloud than a point, and maybe that's why we're
getting such crazy results."

Fine. In that case, let's try the experiment with objects that we may be
certain are objects—discretely localized in space. Figure 11–11a and b show
some photographs of individual atoms. We see for ourselves that these are particles—unequivocally.

Nowadays, individual atoms can even be moved about with "atomic
force tweezers" to make nanoscale constructions. These we may likewise photograph, so we *know* that atoms are discrete objects. After all, they're localized to exactly where we put them, as in the famous picture shown in figure
11–11c.

But if we send atoms through a double slit, we still find an interference
pattern—built up one at a time—as soon as we open a second slit! By behaving
in ways that are sort of like waves and sort of like particles, our atoms manage
to do things that by all rights should be utterly impossible.

BOTH WAVE AND PARTICLE

It's all fine and well to string words together—"atoms are both waves and particles at the same time"—but it's another matter entirely to make sense out of
them. You can't: "I can safely say that nobody understands quantum mechanics," wrote Richard Feynman, a master quantum mechanic, if ever there was
one.[2]

Nonetheless, quantum theory has proven more correct than any other
scientific theory in history. Its predictions, no matter how absurd, have never
been confounded. But this benefit comes at a cost: We can't make sense of it.
The best we can do is to accept the fact that the things which interfere are
purely abstract probability waves, the "only . . . interpretation possible."[3]

Interference patterns formed by many quantum objects are something
like mathematical equations for statistical distributions that nudge otherwise
wholly random individual outcomes into a fuzzy approximation of the equation. "Statistics" are just the rule, or formula, that tells you how various prob-

abilities are arranged. ("What is the probability of the event occurring, or the object being found, at each location in space?" for example.) So, if an interference pattern is formed of statistics, then do probabilities interfere (whatever that means) to produce it?

Almost, but that answer can't be exactly right for a simple reason (and not just because it sounds senseless!). Interference arises only when waves intersect, so that peaks and troughs combine. But the peak (maximum height) of a wave is positive, while the trough (maximum depth) is negative. Bands of zero height (e.g., brightness, were we speaking of light; density of particle strikes with electrons) appear wherever a peak and a trough combine. So, whatever it is that is interfering in quantum events must have the form of a wave and must be able to take on negative values. But, by definition, probabilities proper can only take on values between 0 (a.k.a. "absolutely out of the question") and 1 (a.k.a. "absolutely certain.")

It turns out—for no reason beyond "that's just how it is"—that the "medium" in which quantum waves propagate consists not of probability proper but of "square root of probability," a quantity (both positive and negative) called "probability amplitude." Physicists commonly speak of the waves being of (as in "made of") probability amplitude, but this is not quite precise enough, just as waves of water are really waves propagating in water. That the background medium in which quantum events propagate is something closely related to probability (its square root), thus making "square root of probability" seem like a kind of ethereal "stuff"; and that, upon detection, these waves assume the form of matter (localized as a particle) so that information comes into existence that previously did not even exist (before detection, the "particle" had no location) strongly suggest some kind of connection between matter and information. This particular form of quantum weirdness we will discuss in greater detail later, when we come to quantum computation and its near-miraculous potential for information processing.

In any event, when "square root–of–probability waves" (called "probability amplitudes") intersect, they combine to produce a probability amplitude interference pattern. This pattern has both positive and negative values between −1 and +1. To determine the probability distribution proper, you must everywhere square the values of the probability amplitude.

The most mysterious part of all this, however, is the fact that no matter how precise and mechanically perfect may be the equations that govern the distribution of probabilities for quantum events, the actual event itself—which one of the variously likely or unlikely possibilities that becomes actual and real—is, within these probabilities, completely random. It is "caused" (i.e., selected from among the possibilities) by absolutely nothing in the universe: no force, no collision, no prior events—nothing. It just happens for no reason whatsoever. This is not the seeming randomness of a coin toss. When we flip a coin the outcome is totally determined ahead of time, it's just we aren't interested in and practically speaking couldn't gain access to all the determining causes—the trajectory of our thumb, friction with the coin, air resistance, and so on. But quantum randomness is absolute. "'[C]hance' had for the first time entered the quantum processes. . . . [I]n a sense, a . . . quantum is left to itself to decide when and in what direction it exits." Einstein was fully aware of what this meant: "'Chance' undermines causality and thus topples the framework of classical physics [where] . . . [f]rom a given initial state, a system develops over time with unambiguous regularity, in such a way that all its future states are determined."[4]

It can hardly be exaggerated how disturbing is this quantum portrait of a world where particles go where they will, undetermined. "Anyone who is not shocked by quantum theory has not understood it," said Bohr. That two identical events, set in motion under identical circumstances, will evolve in identical ways to reach the identical end point has been the foundation stone of science, of physics in particular, and "in science there is only physics; all else is stamp-collecting."[5] Has physics given up? According to Feynman, "Yes! Physics has given up. We do not know how to predict what would happen in a given circumstance, and we believe now that it is impossible—that the only thing that can be predicted is the probability of different events. It must be recognized that this is a retrenchment in our earlier ideal of understanding nature. It may be a backward step, but no one has seen a way to avoid it."[6]

This so bothered Einstein that he sacrificed a great part of his later career in a futile attempt to find an alternative: "I would not like to be driven into abandoning strict causality without a great deal more opposition than has been shown so far. The idea that an electron . . . *by its own free decision* chooses the moment and the direction in which it wants to eject is intolerable to me. If that is so, I'd rather be a cobbler or a clerk in a gambling casino than a physicist."[7]

But the only consistent interpretation of quantum mechanical experiments is that nature is inherently probabilistic—that events happen for "reasons" that are not a part of nature. (Scientists prefer to say for "no reason," but that's hardly less mysterious a statement, and entirely indistinguishable from the former in any event.) "Determination," wrote Einstein's friend and colleague Max Born, "must be given up in the atomic world."[8] Einstein, Podolsky, and Rosen, in EPR, showed that, because of the toppling of causality it implied, should quantum mechanics prove a complete description of nature, reality itself, as we experience it, would cease to have the meaning we attribute to it. Having showed that this conclusion follows necessarily, it seemed obvious to them that, therefore, quantum mechanics was inadequate. That's not how it has turned out: "[E]xperiments have conclusively shown that quantum mechanics is indeed correct, and that the EPR argument had relied upon incorrect assumptions."[9]

We are now facing the prospect that there is no completely fixed, objective, local reality and that if not, as EPR argued would necessarily follow, reality is in many respects "nonlocal"—contrary to Einstein's assumptions: Electrons in the double-slit experiment interfere with themselves, it seems. In other setups, they "influence" each other, it seems, without interacting, no matter how far apart they are in either space or time. (See appendix C.)

Furthermore, "wholes" may be composed of "parts" that have, and can have, no causal connection to one another—at least under carefully controlled conditions, such as the double-slit setups: The mere openness of a second slit yields a different "whole"—gun, particles, detector, experimenter. In this configuration, the very nature of reality is somewhat different and so displays to us a different face than when the second slit is closed. Not that opening a second slit causes electrons to behave as waves. There is no way that a cause can be transmitted from the slit to the electrons—especially not to those waiting in the wings to be shot out.

It's easy to understand why so many people—poets, mystics, philosophers, and some scientists too—have seen in quantum mechanics a new religion. But the "reality" of the quantum world is not so simple. It may not fit comfortably into the four-hundred-year-old categories of Cartesian, Enlightenment determinism, but neither does it fit comfortably into any preexisting or freshly minted religio-mystical scheme. It occupies ground never before conceived of—to which we now turn.

12

E UNUS PLURIBUM

OUT OF ONE MYSTERY, MANY

*This theory reminds me of the system of delusions of an
exceedingly intelligent paranoiac, concocted of incoherent
elements of thought. . . . If correct, it signifies the
end of physics as a science.*
—Albert Einstein, letter to D. Liplein, July 5, 1952

It's true that all the mystery of quantum mechanics is contained in the double-slit experiment. But it's a mystery with many facets, and it has taken a century for its many implications to unfold in all their bizarre richness. Nor is this unfolding at an end. Each year, it seems, brings some new and startling revelation rooted in the quantum nature of reality. Quantum teleportation of visible objects; quantum computers whose chips process in a network distributed over multiple universes—these and many an outrage more are no longer the stuff of science fiction but the objectives of serious scientists and engineers at major institutions worldwide: Oxford University, Harvard, MIT, IBM labs, to name a few.

What if anything does all this have to with the brain? Most neuroscientists today insist that the answer is "nothing." And, however tantalizing the analogy, it is certainly true that there is no way of directly and sensibly connecting quantum behavior to human behavior: "Electrons seem to do what they want because they're quantum; must be that I do what my brain wants me to, because it's quantum too" at the best skips a few steps (not to mention facts); at the worst it could well be the "delusion of an exceedingly intelligent paranoiac, concocted of incoherent elements of thought."

But it is possible to indirectly and sensibly connect quantum behavior to human behavior, although making this connection requires considerably more attention to detail than can be encompassed by any poetic analogy. Of the many mysteries that unfold from the double-slit experiment, we will require familiarity with certain key ones in order to build a solid bridge of data and of scientific analysis between the subatomic and the everyday: (1) wave-particle duality, (2) contextuality, (3) indeterminism, (4) absolute chance, (5) lack of trajectories, (6) superposition, and (7) tunneling (teleporting). Then we will take a look at a real-life example of quantum weirdness. But we also need to understand why most biologists doubt that any of this has anything to do with life. Only after we've explored more fully the applications of quantum weirdness now making their way into technology will we be ready to see why those biologists are mistaken.

WAVE-PARTICLE DUALITY

Devised nearly forty years ago, the double-slit experiment brings out the wave-particle duality of quantum reality with great power, as we've seen in last chapter's discussion. Today, however, the wave-particle nature of matter can be seen directly, as in figure 12–1.

CONTEXTUALITY

We have called this "wholeness." The dual-slit experiment forces us to reconsider how we—and apparently nature—define what constitutes a "unit." In the quantum regime, context is an absolutely inseparable aspect of any "entity."

INDETERMINISM

A system of four mutually coupled, perfectly mechanical oscillators can easily give rise to a sequence of states that never repeats itself exactly (chaos). But the states are completely determined by the system's initial state, by the characteristics of each element of the system, and by how these interact with one another. The complicated behavior that arises in such systems is usually referred to as "emergent"; it is not indeterminate.

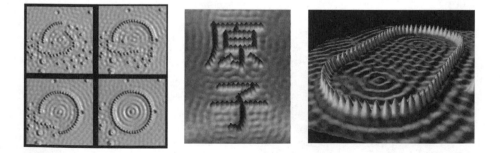

FIGURE 12–1   Atoms arranged by "atomic force tweezers" in various patterns. The wave functions of their electrons (and, to a degree, of the atoms themselves) interfere, producing the waterlike surfaces around the atoms.

The behavior of the double-slit experiment is truly nondeterministic. The density distribution of electron strikes is guided (probabilistically) by an exact equation, but the individual strikes occur at random—the exact locations are determined by nothing that is a part of the universe, yet it seems as if the particles are "cooperating" with each other so as to ensure the proper overall distribution. Such behavior is more than "emergent"; in a truly mechanical universe it would be impossible.

ABSOLUTE CHANCE

The fact that the final locations of particles are indeterministic is another way of saying that there is an element of absolute chance in their outcomes. The eeriness of this idea—how truly difficult it is for our minds to grasp—is nicely highlighted by the fact that ancient and medieval philosophy long ago settled on a definition of God as "the uncaused cause." God is effectual in our material world, in this view, exerting an influence on things, especially in his creation of them, but has himself no cause. The philosophers and scientists of the Enlightenment accepted this definition of God precisely because it explained his absence—there are no uncaused causes in the world they saw; rather "everything happens for sufficient reason." Perhaps God created the world, but it's done entirely without him since. Quantum events, however, as modest as they may be in scale, are precisely "uncaused causes," and they are ubiquitous—hence the suspiciously "religious" character of quantum theory, in Einstein's cool-eyed view.

LACK OF TRAJECTORIES VERSUS . . .

Particles travel along specific, localized paths: an unbroken sequence of points in space, in an unbroken sequence of points in time. Clearly, a "particle" that travels through neither one slit nor the other but (sort of) through both cannot be said to "follow a path" in any commonsense meaning of the phrase. Yet such a quantum particle begins at one definite point in the emitter and ends up at one definite point at the detector. Where was it, then, in between? What was its "path"? (See figure 12–2.)

The fact is that there is *no* possible single path followed by particles in a double-slit experiment. Any reasonable path fails to go through at least one of the slits, therefore eliminating the possibility of a single particle interfering with itself, as it evidently does. And paths must make turns of some sort, which means acceleration, which implies a force, and there is none, thus violating the conservation of energy. Any unreasonable path that has the particle somehow traveling through both slits in sequence (this was at one time a serious, if desperate, proposal) must make some pretty fancy turns, and this means acceleration (as when you speed around a turn in a car), and this means that some force must be acting on the particle to turn it—and no such force is present. In brief, quantum particles do not travel along well-defined paths, even though

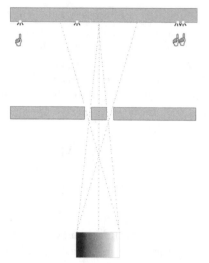

FIGURE 12-2   "Trajectories" in a double-slit arrangement are impossible and nondeterministic. Faint dotted lines show center and extremes of classically possible trajectories, including ricochets, with either one or two slits open. Quantum particles can "tunnel" through barriers.

they do get from here to there—somehow. True, should you place a barrier in front of one of the slits, electrons are apparently stopped there. But remember, a two-slit experiment is a new reality altogether.

. . . SUPERPOSITION

In another manner of speaking (developed by Richard Feynman), we can say that a particle travels via *every conceivable* path concurrently, and it's the paths that interfere. (Now that you've grown used to saying things like "waves of square roots of probability interfere," saying that "paths interfere" should seem far easier to believe, but don't be fooled—no one understands what this "means" either, as Feynman himself would be the first to acknowledge, even though the mathematics work out perfectly.)

Just as we say that two or more water waves combine and interfere—a "superposition" of the constituent waves—and that "probability amplitudes" combine and interfere, we can also say that between the emitter and the detector, electron paths in a double-slit setup are in a superposition. The paths end up weighted probabilistically after they interfere. It just happens to turn out that these paths combine to create patterns of the sort that dictate the bands we saw in the double-slit experiment.

In fact, even if only one slit is open, a single electron traverses every path—including ones that loop through Detroit, take a detour to the Andromeda galaxy, buzz around the local bar, and then head back on target. When you do the math for the one-slit setup, the farther a path is from what common sense expects, the more it interferes destructively with its neighbors, who, being neighbors, are comparably far out; and the closer a path is to what common sense expects, the more it interferes constructively with its neighbors, who, being neighbors, are comparably commonsensical. So it looks to us as if the electron is following a nice, commonsense path. It isn't really, but unless we open a second slit we can't tell. However, like atoms themselves, superpositions can now be directly detected: "A trapped . . . [Beryllium] ion was laser-cooled . . . and then prepared in a superposition of spatially separated coherent . . . states. . . . The . . . superposition was verified by detection of the quantum mechanical interference between the localized wave packets."[1]

TUNNELING

If electrons travel every path, even with only one slit open, what happens if both slits are closed? Could an electron somehow evade the barrier altogether? What if it was trapped inside a three-dimensional barrier—a closed box; could it somehow escape? If everything is merely a matter of probabilities, maybe it could. After all, who says that just because it's "inside" the box (using everyday terms) it isn't also in some sense "outside" it? Roland Omnés, a theoretical physicist at the University of Paris, considers the earth and all its inhabitants thereof:

> If one assumes that the motion of the earth is basically governed by quantum mechanics, one must also consider . . . effects . . . which look incredible from the standpoint of common sense. For instance, nothing forbids the Earth from leaving the neighborhood of the Sun . . . and . . . ending . . . around Sirius . . . [by] a tunnel effect. . . . [T]he . . . probability . . . is . . . [ca.] $10^{-10^{84}}$. However small this probability may be, it is not strictly 0. One can therefore agree with common sense in saying that the Sun will rise tomorrow, but one should allow for the possibility of such crazy event, at least as a matter of principle.[2]

But for an alpha particle (two protons and two neutrons) that form part of a large nucleus, the odds of such an escape are a whole lot better; so much better, in fact, that they allow for a damaging level of radioactivity from certain elements. This is so in spite of the fact that the nucleus forms a cage from which it ought to be impossible for an alpha particle to escape.

Like events happen—in huge numbers—when electricity (electrons) from the wall socket enters the copper prongs of a plug. Your electrician probably doesn't know this, but even the thinnest imaginable layer of copper oxide (tarnish, verdigris) is a fantastically effective insulator. Were it not for massive electron teleportation ("tunneling," as it's called), a copper wire or prong would become perfectly insulated within seconds of being exposed to air. But electrons—like all quantum entities—do teleport: They simply disappear here and reappear there. Since they do not traverse the intervening space, they can escape through barriers that would otherwise be impenetrable. Quantum teleportation is not so far out an idea as it seems—it's going on constantly, everywhere.

MAGIC PYRAMIDS

Now let's look at a real-life example of all this quantum weirdness—the ammonia molecule. It's a good example for our purposes because it's well understood (to the extent that anything quantum can be). Being a molecule it's a lot larger than an electron, so it's somewhat closer in scale to living systems; in fact, it's a molecule commonly found as part of living processes, and so it serves as an introduction to ideas that we will soon be applying to larger biomolecules.

Ammonia is a compound of one nitrogen and three hydrogen atoms—NH_3. It is ubiquitous in both the plant and animal world, as anyone who has ever smelled compost or natural fertilizer knows. The atoms that comprise ammonia form a compact tetrahedral pyramid: One nitrogen atom sits directly above the center of a triangle of hydrogen atoms, as shown in figure 12–3.

Bonds between atoms are more like springs than like sticks. So there is a certain distance above the hydrogen that ammonia "prefers" its nitrogen to be: Too close, and the "spring" pushes it out; too far, and it pulls it in; just so, and the molecule rests in the minimum "basin of attraction" in ammonia's "energy landscape." Figure 12–4 shows this basin.

Depending on the temperature (how much the system is jiggled), ammonia will stay in a configuration that corresponds more or less closely to the minimum energy state, vibrating back and forth around the ideal state. However, there are actually two ideal states, for the nitrogen will be just as comfortable below the hydrogens as above them. So the energy landscape really looks like the graph in figure 12–5a.

This means that if the nitrogen should get to the exact center in the plane of the hydrogens, the smallest jiggle will send it as readily down the energy slope to the one side as to the other. Like an umbrella stretched flat in the wind, ammonia can just as easily snap itself inside out as return to the state it came from.

If ammonia is heated enough, the nitrogen will so readily make it to the center that in the long run, a collection of hot ammonia molecules that all start out with the nitrogen "up" will end up with half of them (at any moment in time) down, as represented by figure 12–5b.

Now, let's chill the ammonia to so low a temperature that the nitrogen can slip through the hydrogens only very rarely because they almost never get

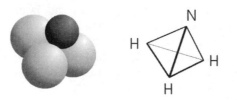

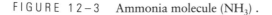

FIGURE 12–3   Ammonia molecule (NH_3).

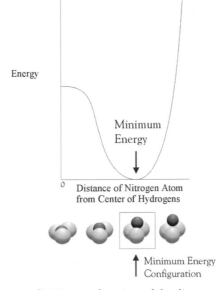

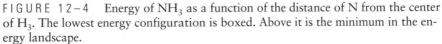

FIGURE 12-4   Energy of NH_3 as a function of the distance of N from the center of H_3. The lowest energy configuration is boxed. Above it is the minimum in the energy landscape.

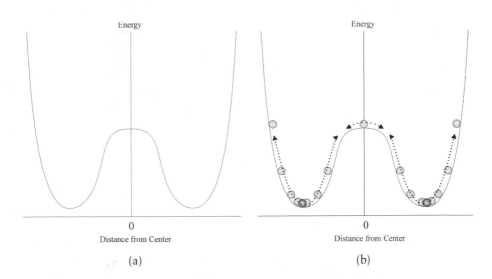

FIGURE 12-5   (a) Symmetrical energy landscape for NH_3. (b) Ammonia's nitrogen atom occupies various positions on both sides of the energy peak.

jiggled enough to reach the midpoint. If all the nitrogens are up, we expect almost all will stay that way. We can arrange it so that it ought to take centuries
for a 50:50 distribution to develop. But in fact, no matter how low the temperature, ammonia molecules are almost immediately found with an equal distribution of outside-out/inside-out configurations. This is because the nitrogen
tunnels through the plane of the hydrogens in spite of the energy barrier, just as
in quantum annealing. The two energy minima are identical, and neither is favored, so both are equally represented. The ammonia molecule executes a complicated, classically impossible "vibration," in which half the time it may be
found near the one minimum, and half the time near the other.* Put differently,
the quantum nature of reality has caused an optimization problem to be solved
in a way that would otherwise be absolutely impossible—ammonia attains and
maintains itself in both of its equal energy minima without anything (like jiggling) making it do so.

AMMONIA SUPERPOSITION: QUANTUM SLOSHING

The two energy minima for ammonia are very favorable. As a result, the molecule is mostly "in" these two states. But more precisely, it is in a superposition
of all possible states with a probability density for every state that corresponds
to the depth of the energy well—the deeper into the well you are, the greater
the probability density at that point. However, in this case, by contrast to that
of the double slit, this probability density function varies not only with distance (in either direction) from the center of the hydrogen triangle; it varies in
time as well. More exactly: The probability density distribution sloshes back
and forth from one side to the other. Why it does this involves subtleties we
need only touch on. But look at it this way: If probabilities can form waves and
interfere, why shouldn't they just as well be able to slosh? In any event, the ammonia molecule is intrinsically quantum dynamic. The superposition is built up
out of different proportions of every possible state; and these proportions vary
rhythmically in time. The sloshing does not mean, therefore, that the molecule
actually goes back and forth between its two low-energy states, so that it's in
one at one time and in another later. It's always superposed in both. The probability of catching it, however, does indeed slosh.

We can also put it this way: The up-and-down energy regions of an ammonia molecule are like two separated basins connected at their bottoms by a
U-shaped hose. (Remember the quantum annealing energy tunnels.) Without
the hose, if one basin is filled with water and the other is empty, that's how the
system will stay. Add the hose, and the water sloshes back and forth from one
to the other. If there's no friction, it'll slosh forever. Think of there being
enough water so that there's always some water everywhere, but which location has the most water travels to and fro. Then the amount of water at any location represents the probability of finding the nitrogen there. The probability

---

*
 We can only say "*may be found* near." We may not say "is near." Before we detect it,
the nitrogen "is" both everywhere and nowhere, with a range of "probability amplitude densities."

itself obeys nice, neat mathematical rules; the finding of a nitrogen of one single amonia molecule at an actual location obeys no rules whatsoever. As Max Born put it: "The motion of particles follows probability laws, but probability itself propagates in conformance with the law of causality."[3]

MANY SHALL RUN TO AND FRO (AND KNOWLEDGE SHALL BE INCREASED)—DANIEL 12:4

It is not at all easy to speak of the nitrogen's position and not be in some way wrong. But let's try. We say, "The sloshing wave equations provide us with a(n ever-changing) picture of the probability of finding the nitrogen at any point at or in between the up and down extremes." Under normal (classical) conditions, this would also mean, "The nitrogen is at one of these points—we just don't know which one; we're only able to assign a probability to each point. The best guess as to where it actually is—like the bead under some shell in a shell game—is, of course, the point with the greatest probability."

But in the quantum world the latter statement would be false. The first statement says that there is "a probability of finding the nitrogen" at any given point—and this is literally true. If we actually do something to try to find it—lift all the shells in a shell game, as it were—it will in fact be seen at a particular spot, just like the electrons in the double-slit experiment that came to rest in a particular location. But before we actually detect its position, the nitrogen atom has no fixed position; it is in a weighted distribution of all possible positions—a superposition.

GOOD VIBRATIONS

In spite of the utter and absolute (if nonetheless probabilistically weighted) randomness and indeterminacy of where we would find the nitrogen were we to look, when we aren't looking, the probability distribution sloshes back and forth with incredibly mechanical precision, giving the ammonia molecule a permanent natural frequency of quantum "vibration": 2.3786×10^{10} cycles per second. This frequency is so fast and so precise that it was long used as the basis for high-tech clocks. (Now other, faster quantum vibrations have been put to the same use.)

Most of the tunneling effects long known to physicists—such as those we've discussed in this chapter—occur on very small scales and take place in biologically inert situations. Where these effects have been exploited and amplified in scale for our use—as in the development of electronic materials, for example—it is only by applying extraordinarily intensive technologies under fiercely controlled conditions.

This, in a nutshell, is why most serious scientists doubt that quantum phenomena will ever prove to be a significant part of our understanding of anything living, let alone anything as relatively huge and complicated as the brain. From the classical world, with its mechanical determinism and lack of freedom, quantum phenomena cannot be expected to provide an escape. To understand the subtle ways in which this is incorrect, we need first to penetrate deeper into quantum mysteriousness and its applications—in particular, we need to grasp something of the peculiar connection that quantum mysteriousness reveals between matter and information. We will then be prepared to descend even more deeply into the structure of life, and especially of the brain, a material entity that nature has designed most exquisitely, it seems, to process information.

13

STRANGER IN A STRANGE LAND

(THE STRANGER THE BETTER)

Sanity, I submit, is not a canon of science.
—Julian Schwinger, *A Progress Report: Energy Transfer in Cold Fusion and Sonoluminescence* (1991) (Schwinger shared the 1965 Nobel Prize for the development of quantum electrodynamics with Sin-Itiro Tomonaga and Richard Feynman)

Physicists have had a very hard time swallowing the bitter pill of absolute chance but at long last have done so. However, this is only the first pill of many. There is far more to quantum weirdness than only absolute chance—and far more that is useful. Let's now look more carefully at some of its strange effects and at how these effects can be implemented so as to generate astounding intelligence—and even a capacity for a kind of intelligence that seems not so much human as superhuman. We will then be in a position to see what, if any, relation these strange effects have to brain or mind. The key link among all these effects is the peculiar and intimate relationship between quantum phenomena and information. We will discover that if the matter is essentially quantum, then matter is information.

MAXIMUM INFORMATION

Imagine a shell game with just two shells. The guy who's scrambled them isn't conning you—he's just fast. There are only two possible outcomes, and only one is the case: Either the left shell has the bead (a state we call "L," as in "L is the case"), or the right shell does (a state we call "R"). The fact of either L or R constitutes one binary bit of information. We could think of the state of the shell game as like "the state of a cell in a cellular automaton"—one "cell" with two possible states. We could then think of the state of your ignorance of whether L is the case or R is the case as a state of 0 bits of information at "your" location—an adjacent cell. When we learn where the bead is—whether it's R or L—one bit of information is copied from the shell cell to your cell. At present, both shells sit in the closed position and you have 0 bits of information.

You place your bet and lift a shell. If the game is fair, then whether the bead is there or not, whether the outcome is R or L, whether you win or lose, you acquire 1 binary bit of information: you now know with 100 percent certainty where the bead is. In real life you'd want to confirm this by lifting the second shell—if you can.

The guy scrambles the shells once more. Once again you have 0 bits of information. Your single bit has been erased. Each time a new piece of information is preceded by erasure of the old. Where the bead is now tells you nothing about where it was on any preceding scramble.

Now let's play a quantum shell game. For this, we could use any one of a number of binary setups. For example, we could use the two energy minima of the ammonia molecule, or the up versus down spin states of an electron or of an atomic nucleus. But the easiest one to understand and visualize (sort of!) involves photons of light. It's a model that actually has been implemented, and we can apply to it some of our earlier discussion of the double-slit experiment.

Instead of shooting a beam of light toward a barrier with openings in it, let's shoot the beam at a device called a "beam splitter." This is simply a very high-grade half-silvered glass: a "one-way mirror." When properly angled, half the light is transmitted through the glass and half is reflected, as in figure 13–1.

We're also going to control things tightly. First, all the light we use—whether many photons at a time or few—comes from a laser, so it's coherent (waves all in sync). Second, there are no irregularities of any sort (like bumps) in the splitter that can cause certain photons to behave differently from others. If we think of the incoming light as a pure, continuous wave, then all the splitter does is to allow half the "brightness" to pass straight through and the remaining half to reflect off at 90 degrees.

But matters aren't so simple when we consider that the beam of light is composed of individual particles—photons. Then the fact that this is a nearly

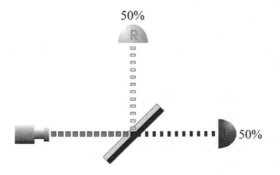

50%

50%

FIGURE 13–1   A beam-splitter. Half the light is reflected (R—directed upward) and half transmitted (T—allowed to pass through).

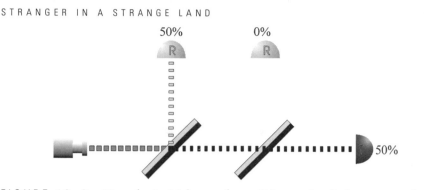

FIGURE 13–2   Hypothesis: "A beam of pure T (transmitted) photons must be 100 percent transmitted through a second identical beam splitter." *Right or wrong?*

perfect beam splitter, combined with the fact that exactly half the photon count appears at one detector and half at the other, makes us wonder whether there are two types of photons, those prone to transmit (T types) and those prone to reflect (R types). The statistics show an overall 50:50 division of R and T, but the order in which they emerge is completely random—like the sequence of boy babies and girl babies born in a hospital.

So, perhaps the splitter is a kind of filter—it's not actively affecting truly identical particles but rather just passively sorting subtly different ones. Since all the R photons have been directed upward and all the T photons forward, we make the following prediction: If we place a second identical beam splitter in line with the first, that second splitter will be struck exclusively by T photons. All of these photons, therefore, should be transmitted through the second splitter, with none reflected, as in figure 13–2.

Consider the first arrangement, with only one splitter. The laser is like a shell shuffler. We have no way of knowing the type of an emerging photon until we raise a shell—that is, see which detector registers the photon's strike. If it's the R detector, we know that the photon was an R-type; if not, we know with 100 percent certainty that it's a T-photon and that the T detector will register it. We don't really need the T detector; it's doing no more than raising the second shell to confirm that's where the bead is, if it wasn't under the first. If we do use it, it always confirms what we already know; it adds no new information, just as lifting the second shell adds none. We get 1 binary bit of information regardless of what happens.

In this spirit we add the second splitter. Everything else remains the same. The second beam splitter just delays when we lift the second shell for confirmation. No new information can be obtained this way either. One bit—that's all we get. This makes sense. After all, one photon coming through the setup itself carries only 1 binary bit of information—it's either an R or a T—and we uncover that information at the first beam splitter. There's no more information produced by the second beam splitter because we've extracted all there is at the first. So our commonsense hypothesis tells us.

This is not what happens—not even if we send photons through one at a time and in exactly the same location. No matter what pains we take to ensure

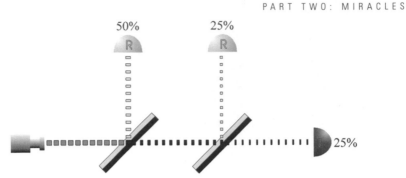

FIGURE 13-3   Reality: After passing through a beam splitter that allows only T-photons to pass, a second identical beam splitter somehow re-randomizes the T-photons into equal proportions of T and R.

that we are dealing with an absolutely pure, coherent beam of identical T photons, when we send that pure T beam through a second beam splitter, the beam again sorts the photons at random into T photons and R photons—in spite of there being nothing in the splitter to cause this to happen. (See figure 13–3.)

We're back to "How can it be?" We know that the sorting isn't due to something in the splitter, so it must be something in the photons. But now we've proven that it's not something in the photons, so it must be something in the splitter.

Here's how weird this is. As babies are born at a local hospital, the girls are brought through one door into a pink room, the boys through another door into a completely separate blue room. But when the babies are brought out of the pink room, or out of the blue room, through doors at the other end, they once again must be sorted into boys and girls.

There is an "answer" to this dilemma that works for photons and other quantum entities, but not for anything else: T photons aren't turning into R photons and vice versa because they aren't simply T or R to begin with. After passing through the first splitter, photons are in a superposition of R and T states. Only when measured, after striking one of the two detectors—not before!—do they become exclusively R or T, and which of the two any single photon ends up as occurs absolutely at random, but with long-term odds of exactly 50:50. The act of detecting a photon forces it to "choose" either R or T. Before that it is both and neither.

There are some less than self-evident informational consequences to this most strange state of affairs. For one thing, if you keep sending a mixture of (supposedly) R and T photons through a series of beam splitters all oriented along the T direction, you'll "convert" an ever larger total percentage of the photons into R types. (See figure 13–4.)

But, actually, the converse of this is of greater interest. Consider a single photon reaching the first in the line of splitters. It has a 50:50 chance of registering as either an R type or a T type. Suppose it's an R type. It gets reflected and registers a strike at R Detector #1 (on the left in figure 13–4). Suppose it's

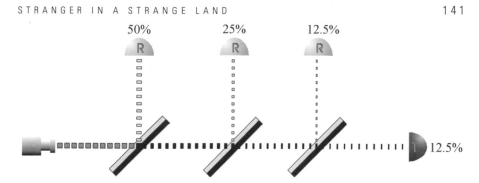

FIGURE 13–4   A sequence of beam splitters converts a 50:50 mixture of R and T photons into an arbitrarily large percentage of R beams.

a T type. Then the way things ought to work is that the T-type photon will be transmitted through every beam splitter and will eventually strike the T detector on the far right at the end of the line (which we may make as long as we like). Therefore, since no R detector would ever register a strike, no R detector other than #1 could give us any additional information.

But that's not how it works at all. Instead, there's a 50:50 chance that the photon is an R type, and if so, it strikes R-Detector #1. *We thereby retrieve the 1 binary bit of information carried by the photon.* That is, we learn something we didn't know before—which of the two types (hence, "binary" information) the photon is. If it doesn't strike R Detector #1 and proceeds to beam splitter #2 (still carrying its single bit of information), there is suddenly, once again, a 50:50 chance that it will be reflected or transmitted. If it is reflected, we'll detect it when it strikes R Detector #2. But that means that we know that it was transmitted at beam splitter # 1—1 bit of information—and reflected at beam splitter #2—*a second bit of information.*

We may place as many beam splitters as we wish in this line. Which detector any one photon will strike is completely random (within certain probabilities in the aggregate), and there is some possibility that the photon could strike any detector at all, no matter how far down the line. The detector that registers the strike reveals the unique sequence of reflections and transmissions that preceded it. Instead of just R or T—a single binary bit—we can get, say, TTTTT (five 50:50 opportunities in a row translates into 2^5 = 32 bits), or TTTTR (also 2^5 = 32 bits), or TTTTTTT (2^7 = 128 bits). If we add a beam splitter after every split (in figure 13–4 above, imagine lines of splitters running vertically as well), we can get every possible combination of R and T, without limit.

Thus, out of the rule-constrained but absolute randomness of this quantum system there seems to be the potential for an unlimited number of bits of information. *Absolute chance therefore implies infinite information.* Each time a wholly uncaused quantum event comes into existence, it's a new piece of information in the universe that was not there previously; it wasn't in the least implied by a chain of prior events.

Back on the No-Path

From another point of view, we're back to quantum particles having no path at all and yet following every path at once. The beam or the single photon is not being sorted into R or T types. It therefore can't be said to be traveling along one trajectory or the other—vertical for R, horizontal for T. As it leaves the splitter, the beam (or the single photon) is in a superposition of *both* R and T, just as once both slits are open the photon in the double-slit experiment is in a superposition. It is this superposition that carries the extra informational bits, with the potential for an infinite number of bits.

Now, suppose you object: "Superposition? How do you know that? For example, I don't see any evidence of interference." There's no interference because after the photon strikes the first splitter, its (superposed) components are directed away from each other. To make them interfere, we would need to redirect the half beams so they'll intersect. Let's do this by adding two ordinary mirrors to our setup. But let's also reposition the second beam splitter and add a second source of coherent photons to create the arrangement of figure 13–5. By "coherent," we mean that the two sources are both lasers that therefore produce light (photons) of but a single, pure frequency with waves and troughs for every photon perfectly aligned (in phase). Furthermore, we arrange matters so that the peaks and troughs of the light (photons) from each laser are aligned with the peaks and troughs of the light from the other.

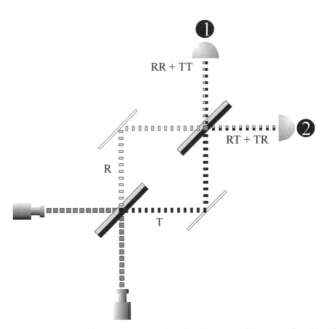

FIGURE 13–5   A second beam splitter that both recombines and splits the already split beams. To get to Detector #1, a photon can either be reflected twice or be transmitted twice (RR or TT). To get to Detector #2, it must be reflected once and transmitted once (RT or TR). Two quarter-height waves combine at each detector as shown. (We are not counting reflections at either of the two regular mirrors.)

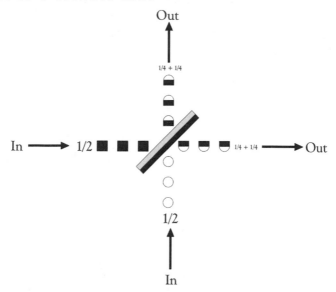

FIGURE 13–6   Detail of second beam splitter now handling two sources of photons.

The second beam splitter now does two things at once. It both splits the two incoming half beams into quarters, and it recombines them into two new half beams, as shown in figure 13–6.

That there is in fact a superposition and that even *a single photon coproduced by the two sources* therefore has the capacity to interfere with itself can be brought out as follows.* We ask, "Are the intensities of the beams emerging from the second splitter equal?" In this exact arrangement, they are. But there is something different about the second splitter in this arrangement that we haven't mentioned yet: One of the two half beams hits it from the back.

This is important because the beams of particles are also waves. To make the recombined quarter beams all come out right at the end with respect to their phases, the second beam splitter has to transmit without change the wave that passes through from front to back but *flip over* the wave that passes through from back to front. That's the same as shifting it out of phase by 180 degrees—its peaks become troughs and its troughs peaks. (See figure 13–7.)

This creates a most interesting opportunity. If we move one of the regular mirrors just a tad, we can lengthen one of the paths and create a phase difference of any proportion we wish among the various quarter beams. Depending on the difference between upper and lower paths, we can make the two vertically directed quarter beams combine constructively and the horizontally directed ones destructively, or vice versa—or anything in between.

---

* Appendix C explains what the italicized phrase means and discusses the famous Pfleegor-Mandel experiment that demonstrated it. In brief: Not only do photons travel along all trajectories at once, the point of origin of the trajectory can be a superposition, as well. For more details on the preceding discussion, see *The Feynman Processor,* by Gerard J. Milburn (Reading, Mass: Perseus Books, 1998).

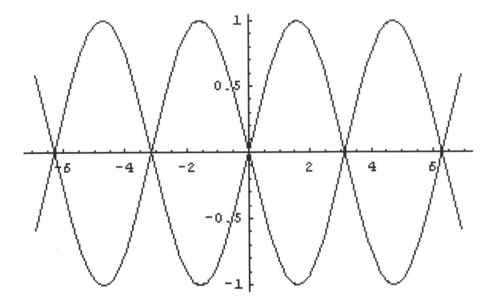

FIGURE 13–7   Two waves 180 degrees out of phase, the maximum possible. If two such waves are added, they cancel out each other perfectly (100 percent destructive interference). At 360 degrees out of phase, they are perfectly in phase again (100 percent constructive interference).

In particular, we can arrange things so that the intensities of light measured at the two detectors vary inversely with one another. At one extreme, every photon hits one detector, and none hits the other. At the other extreme, it's the reverse. At the midpoint, it's 50:50, as before. Only interference can explain this. In a way, it's a fancy variation of the double-slit experiment, except this time the "slits" aren't all-or-nothing affairs; their "penetrability," as it were, is adjustable.

Here's the important point that this adjustable arrangement highlights: A photon is in a superposition of all paths *until it strikes one of the two detectors.* When it does, the superposition collapses. Technically, the superposition includes even wild paths that wander all over the universe. But because of the physical arrangement of the apparatus—a distinct "whole," with qualities that are more than the mere sum of its parts—the wild paths all interfere destructively. Only two are left with any significant probabilities, each of which can be attained by two different routes: RR or TT → Detector #1; RT or TR → Detector #2. Now, just as in the double-slit experiment, there is no way of determining which path any individual photon takes without intruding upon and in some way altering the "whole." More precisely, detecting "which path" forces the photon to choose a path, and that's the one we detect. By moving the mirror just so, we may adjust at will the relative fraction of detections for each path in a large number of cases. But, as usual with matters quantum, for any single detection event, the outcome is absolutely random.

When such a situation gets established between two paths (or more), we deem the paths "indistinguishable." When two paths are indistinguishable, the final distribution of probabilities for the two outcomes is determined by "probability amplitudes," a purely quantum wave phenomenon that allows for interference.

Control over interfering quantum probability amplitudes leads directly to certain strange and rather amazing applications: quantum computation, for example, in which a parallel processing network is created by—suspend your disbelief for a moment—a single quantum processing element with copies of itself distributed across multiple universes, all interconnected.*

THE REST OF ALL POSSIBLE WORLDS

Let's set the path length so that all photons arrive at Detector #1. If we do something to learn which of the two paths (RR or TT) a photon takes to get there—for example, by inserting yet another detector in its path at location #3 (see figure 13–8), we intrude on the superposition and the interference disappears. Moving the mirror would no longer create a reciprocal variation in intensity (frequency of detection) at Detectors #1 and #2. We thereby force the photon to "choose" one of the two paths that diverge from the first beam splitter. (We could just as well have something ricochet off the R photons—another photon, for example. Even if such a photon made it to Detector #1 anyway, it would no longer have the same phase relation. This would destroy the coherence, hence also the superposition.)

But look carefully at what this means: a "which-path" detector at #3 destroys the superposition *whether the photon is in fact detected at #3 or not.* Half the time the photon won't go to #3 anyway but will instead head toward the second splitter. Common sense tells us that for any photons transmitted at the first beam splitter, whether there is or isn't a detector at #3 should be irrelevant. It's not. For now the position of the upper mirror is irrelevant. With no interference and no recombination of quarters, the second splitter simply splits the incoming beam. Result: Half the photons now are detected at #1 and half at #2. An absolutely certain final outcome—every photon is detected at #1 when Detector #3 is absent—is converted into an absolutely uncertain one—some photons are detected at #1, some at #2, and some at #3 when Detector #3 is present.

Many such beam-splitting arrangements (let's call them "quantum 2-bit gates") may be strung together to form networks of any architecture you wish. For example, you chain two gates one after the other, then instead of two indistinguishable and superposed paths leading to the output, there are four. If you chain three gates, there are eight paths. In general, n gates creates a superposi-

---

* Parallel processing using mulitple, superposed states of some single thing has already been accomplished. The interpretation of a superposition as being a temporary connection among versions of that same thing in multiple universes is accepted by only a minority of serious researchers in the field. But among them are some of the founders—David Deutsch, for example, at Oxford.

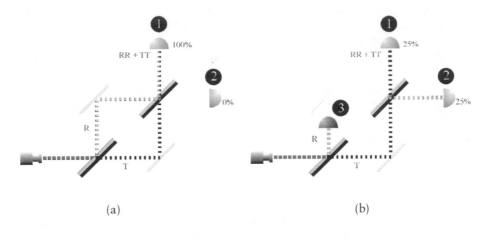

(a)   (b)

FIGURE 13-8   Learning "which path" destroys superposition. (a) Two beam splitters with path lengths arranged so that all photons are detected at #1. The paths that
form the central "square" are in a superposition. (b) An extra detector, #3, forces 50
percent of the initial photons onto the R path. This destroys the superposition and
causes the remaining 50 percent to be distributed randomly between the two final detectors.

tion of 2^n indistinguishable paths. You can also connect gates so that outputs
from one gate are distributed to two and then four and so on, the way that
neurons in parallel-processing neural nets distribute their output. The output
from more than one gate can combine as the input to a third. In this way, gates
may be strung together both in parallel and in series, yielding networks very
like neural networks. But with a quantum network, a given data bit does not
follow a single path (or a limited subset of paths)—it follows every possible
path. A complex input to many gates at the front end then will not be processed according to a single embedded rule (like a fixed set of connection
weights); it will be processed according to all possible rules. It's as if all possible variations of HER response to our initial move in a game of Hexapawn
were to happen at once—so long as there is no intrusion from the environment
to destroy the superposition.

After all the quantum processing has occurred, we do indeed intrude: We
deliberately destroy the superposition in the act of detecting (forcing) a single
end result.

Do-Be-Be-Be-Do

The basic idea behind a quantum computer, then, is that data can be entered
and placed into a superposition which transforms a certain number of input
bits into an arbitrarily large number of superposed quantum bits upon which
computation can occur simultaneously. For example, suppose we have four input bits—say, 1001. As these classical bits enter four beam splitters at the front
end of such a device, they are converted into a superposition of all possible

combinations of four-digit binary numbers: 0000, 0001, 0010, 0011, 0100, 0101, 0110, 0111, 1000, 1001, 1010, 1011, 1100, 1101, 1110, and 1111—because each "either 0 or 1" bit has been converted into a quantum bit, or "qubit."

So, "where" are these sixteen different numbers stored during the processing phase? In a normal computer, sixteen different addresses representing sixteen different physical locations are required to store them. In a quantum computer, all sixteen numbers are stored in the same place. It's like the palimpsest of superimposed memories in a neural net, but of an even stranger kind. For now, the sixteen numbers are not encoded as a fixed configuration of weights, each weight in a different place, all linked to and affecting one another via the intervening processing elements (neurons). The "quantum network" that stores these numbers is a superposition. But where is that located? In what actual location in the universe, and in what physical medium are these numbers encoded?

David Deutsch, the Oxford physicist who first showed that quantum computation could really work, would quite seriously answer that, at the moment of superposition, there comes into existence an additional fifteen universes. During the computation, we cannot tell which of the sixteen is "ours," because, for one thing, to tell, we would need to insert a detector, and that would destroy the coherence, the superposition, and the computation. But that's not all. During the superposition, the sixteen numbers aren't assigned to the sixteen universes, one to each. None is assigned to any and all are assigned to every. Only when we "ask" for an answer, by inserting an external probe, does the assignment get made, and that's the answer we obtain. Who or what does the assigning? Nobody knows. We do know, however, that it's nobody and nothing that is part of our physical universe. In any event, once we do probe for an answer, the "cross-talk" between universes (the superposition) collapses.

Quantum computation is therefore like parallel processing in many networked universes at once. Or it's like a cellular automaton in which each cell exists in many different states simultaneously, in many different universes. It's the universes, in this view, that exist in parallel; we merely take advantage of this phenomenon and get away with performing a four-bit computation on a two-bit device that we need build in only one universe—ours. Now, with this general background, there is a relatively simple, dramatic way of grasping the sheer magic of quantum computation and its mysterious relationship to information (or, if you will, "knowledge").

ZEN AND THE ART OF QUANTUM COMPUTATION

The man stands up on stage, a heavy black blindfold wrapped tightly around his head, blinding him. His polyester sharkskin fabric suit glistens in the stage lights. A woman in the audience holds out her fist. It is closed tightly around a coin—or maybe not. The man bends, tenting his fingers on his forehead in apparent concentration. The crowd is hushed. After a few moments, he shouts: "Yes!" The woman opens her hand. The coin is there and the crowd erupts in

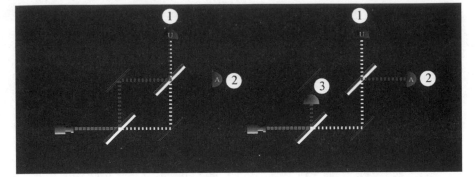

FIGURE 13-9   Double beam splitter setup. (left) without and (right) with an extra detector at #3.

tumultuous applause! "No," he admits, feigning modesty so as to seem more convincing, "I can't be right every time," but he's been right seventy-nine times out of one hundred—nearly 80 percent—far better than chance expectation, as the illuminated record board shows in brilliant flashing lights. The man is a genuine clairvoyant!

Nah. He's just an old-fashioned con man. But what con men have been faking for hundreds of years it appears that quantum mechanists are almost doing for real. Figure 13–9 shows our dual beam splitter again. It's in reverse black and white because we're going to conduct an experiment in the dark: We won't be able to tell whether our gate is as shown on the left, with only two detectors, or as on the right, with a third. To repeat: Sans the third detector, an incoming photon will be placed in a superposition by the first splitter and then recombined and resplit by the second.

Once again, the path lengths have been adjusted so that any photon is detected at #1 with 100 percent certainty. But with Detector #3 in place—or any kind of barrier—the superposition (closed square path on the left) is destroyed and any photons not trapped by Detector #3 can now distribute themselves equally at both Detector #1 and Detector #2.

Now note: A photon can appear at Detector #2 only when Detector #3 is there *and* the photon doesn't hit it. In other words, if we fire but one photon, and it happens to hit Detector #2, *then we know that Detector #3 is there even though we have interacted with it in no way whatsoever*. If the import of this seems unclear, consider the following:

Suppose Detector #3 is actually a sensitive photoelectric trigger wired to a bomb—so sensitive that even one photon striking it would blow you and your lab sky high. The detector and bomb may be there or may not be—you don't know, and you can't even open the door to see, because to see it, light would have to fall on it and reflect back to you, and that would set it off. But you must determine whether it's there or not. Suppose reality was what most people think it is—no weirdness, quantum or otherwise. Suppose, too, that the bomb really is there (but you don't know whether it is or not). Then the cost of finding this fact out—that is, of acquiring 1 binary bit of data: the "yes," of the

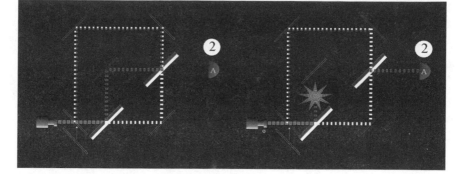

FIGURE 13-10   Quantum bomb detector with a single photon. (left) No bomb at #3. The photon is fed back into the system. (right) Bomb at #3. The photon is either detected at #2 or caught by the bomb's trigger.

two possibilities "yes" or "no"—is a 100 percent chance of detonation. In short, the chance of actually detecting a bomb safely is zero.

But with quantum weirdness, and if the bomb is there, two photons out of four will strike Detector #3 and blow you up, two will avoid it and reach the second beam splitter. Of these, one photon will hit Detector #1, which would fool you into thinking there was no bomb, but one photon out of four will arrive at detector #2, letting you know that the bomb is there without ever interacting with it. (No surprise that this scenario was dreamed up by two Israeli physicists.)[1] The odds of a safe detection of an actual bomb have gone from 0 percent to 25 percent. You've acquired one bit of information "for nothing," like clairvoyance, and you have only one chance in four of succeeding—so call it "a quarter bit." But it's a very valuable piece of information considering what it costs otherwise.

But we can do better. Recall how, as a single photon passes through a line of beam splitters, a potentially unlimited amount of new information appears. With this in mind, we replace Detector #1 with a set of mirrors that redirect the same photon back into the system, again and again, as in figure 13-10. (The last mirror slides into place before the photon makes the first full loop.) If there's no bomb, the photon circulates indefinitely (see figure 13-10 left). If there is a bomb, the photon eventually either detonates it or is detected at #2 (see figure 13-10 right).

In 1,024 trials, 512 photons detonate the bomb and 512 don't. Of the latter, 256 are safely detected at #2 and 256 are recirculated. The same proportions recur with recirculated photons as detailed in table 13-1. $41/(682+341) = 1/3$, so now, when the bomb is there, the odds of a safe determination increase to 33 percent.

If the beam splitters' transmission/reflection ratios are properly adjusted, these odds can be made as close to 50 percent as we like. Paul Kwiat and colleagues at Los Alamos have shown that if a second quantum dichotomy is made in the split beams (e.g., by polarization filters), safe determination rates

1st or Recirculated	Explode	Safe	Detected at #2
1024	512	512	256
256	128	128	64
64	32	32	16
16	8	8	4
4	2	2	1
Totals	682		341

TABLE 13−1   Recirculated photon in a quantum bomb detection scheme.

can theoretically approach 100 percent. Recently a Swedish research group attained an actual 80 percent detection rate.[2]

This technique goes beyond simple yes/no detection. Kwiat and his colleagues have used it to measure the thickness of a human hair, a knife-edge, and a thin wire.[3] Kwiat comments: "The next logical step for us is to use this higher-efficiency system to produce one-dimensional images of objects. If you can do that, you can also produce 2D images—it's not really any more difficult."[4]

"This would open the door, for example, to medical imaging that 'sees' the internal structure of the body without actually sending any kind of radiation—or anything else—into it."[5]

From PSAT to QSAT

The first full-fledged act of quantum computation was announced in April 1998, at least ten years before almost anyone had predicted it would be possible. The problem it solved was, in form, much like math SAT questions, as shown in table 13–2.

To answer, you must actually perform the calculations, one at a time. (If that's not evident from question 1, it surely is from 2.) On average, how many error-free calculations must be done? The sequence in which we test the equa-

Instructions: Fill in the circle next to the equation to which the given answer is the correct solution.

1. X = 3	2. X ≈ 17
○ (a) $4 + 1 = x$	○ (a) $^{1.99}\sqrt{1.000631477 \times 321}$
○ (b) $4 = 3 - x$	● (b) $^{2.00}\sqrt{1.003258657 \times 293}$
● (c) $x = 4 - 1$	○ (c) $^{2.10}\sqrt{1.002165351 \times 338}$
○ (d) $x + 1 = 3$	○ (d) $^{1.90}\sqrt{1.003359821 \times 272}$

TABLE 13−2   Multiple-choice calculation question.

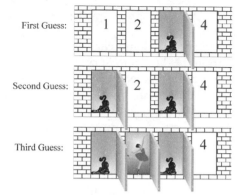

First Guess:

Second Guess:

Third Guess:

FIGURE 13–11   Lady and Three Tigers. It requires on average between two and three guesses (2.2) to choose the Lady. If a wrong guess is fatal, the odds are 3 out of 4 that you'll die on the first guess.

tions has no effect on the long-term statistics. Thus, there is one chance in four that we will select the correct one on the first try. The expected number of calculations that we'll need lies between 2 and 3—2.2 to be exact.

The challenge of selecting the correct calculation to test is a modified Lady and Tiger situation—one with three tigers—in which the odds of getting eaten on the first guess are 75 percent, as in figure 13–11.

This is exactly the kind of situation where clairvoyance would come in handy. Using quantum computation, then, Isaac Chuang at IBM Almaden Labs, Neil Gershenfeld at MIT, and Mark Kubinec at Berkeley did something very close to just that, reducing the correct selection of which equation to test to a single evaluation, occasionally two.[6] In other words, nine times out of ten they chose the Lady with but a single act of choosing. How did they do it?

Like photon direction or polarization, or the spin of electrons, nuclear spin, too, is a quantum phenomenon and can be manipulated to form qubits—multiple superposed binary bits—with the spin-splitting counterpart of a beam splitter: a magnetic field. They placed a jarful of chloroform molecules in the magnetic field of a standard piece of medical magnetic resonance imaging (MRI) equipment and so forced the spins of carbon and hydrogen nuclei into a useful state of superposition. The entire "computer" fit on a tabletop and could calculate all four equations simultaneously—by doing one in each of four parallel universes, if you like. It could therefore tell behind which door was the Lady on the first attempt. More precisely, the average was not exactly 1, but 1.1—it got eaten now and then, but only one time out of ten, instead of three times out of four.

Using related quantum weirdness in a different arrangement (closely related to the quantum computation method described here), it would be possible to construct a device to measure blood sugar levels from outside the body. A tabletop-size prototype of the underlying technology is currently under construction at the "Things That Think Consortium" at the MIT Media Laboratory.

Wu Wei: Being as Doing

At the first NASA International Conference on Quantum Computation and Quantum Communication, held in Palm Springs in February of 1998, Richard Jozsa, of the Department of Computer Science at the University of Bristol, explained that "[T]he act of 'doing nothing' . . . is the essential ingredient [in] . . . quantum algorithms. . . . [S]uppose that we have a quantum computer with an on/off switch. . . . Then (in certain circumstances) the mere fact that the computer would have given the answer if it were run, is enough for us to learn the answer, *even though the computer is in fact not run.*"[7]

The method behind such a Zen computer—a "counterfactual" computer as it's called—follows directly from the bomb-testing scheme. The specifics of how to "perform" a computation without actually performing it are a bit subtler than merely seeing something without looking, but they follow directly from the same principles.[8] In any event, it is almost irresistible to imagine that there must therefore be some connection between that in us which "knows" and "creates"—that is, our minds—and the eerie capacity of matter, under the right circumstances, to function as a seeming generator of information that previously did not exist. Since the material partner of the human mind is the human brain, is the matter of the brain somehow responsible for mind because it, too, functions like a biological quantum computer, just as it functions as a biological neural net? Roger Penrose, a mathematical physicist at Oxford, thinks that such a direct link between mind and quantum is not only possible, but likely.[9]

However, in order for an entity to function as a quantum computer, it must be perfectly isolated from the environment. Quantum computational devices are extraordinarily artificial, depending on settings that are extraordinarily well controlled. This kind of computation depends on the absolute randomness that undergirds all of reality, it seems, but it is utterly and immediately destroyed by even the tiniest intrusion of pseudorandomness—for example, heat—forced upon it from the outside. There is no evidence that quantum computation—of the above sort—is even remotely plausible in living systems at the cellular level where Penrose believes it should reside.

The Golem

There are, however, two ironies. First, there is *another* kind of quantum computation that the human brain may very well employ. We confront it at the next scale down in our exploration of the brain.

Second, while the human brain may not be a quantum computer of the first kind, there is every likelihood that we will soon set evolving on their own devices that are. After all, we've already seen how researchers have created hardware that evolved an unexpected—and still unexplained—use of (presumably) classical electromagnetic effects. (See chapter 8.)[10]

Let me put strongly a hypothesis: *Brains that are quantum computers of the second kind were required to bring into existence brains that are quantum computers of the first kind.* Whether this will prove to our advantage or detriment is a matter of no small concern to those who are on the forefront of their creation. In the words of Kevin Warwick at the University of Reading in Great Britain, professor of cybernetics: "I can't see any reason why machines will not be more intelligent than humans in the next 20 to 30 years and that is an enormous threat. . . . We're talking . . . [of] the end of the human race as we know it."[11]

Expert systems are ever more successfully mimicking the key features of what we once thought inimitably human. We coolly look past the most self-evident characteristic of *emotion*—how it "feels" (to us)—and without sentiment ask instead what role it must serve in a self-organizing computational entity, so we may enhance the artificial brains we create. The answer: Emotion acts like a multidimensional scheme of inter*personal*—by contrast to inter*neuronal*—"connection strengths." The spectrum of preferences associated with each emotion, embedded in each individual "processing element," allows the system as a whole (the neighborhood, the community, the society) to generate a self-regulating social order. Codify these hierarchies, encode them in software, use the software to drive a robot, and the result is an artificial system that displays—with eerie naturalness—the social interactions of a human child. Allow the code to adapt to those who interact with the robot, and the robot undergoes maturation and displays characteristics once thought uniquely human. This has already been done at MIT.

With such capacities already being implemented, once the *hardware* is made both adaptive and small enough to encounter and take advantage of quantum effects, it's difficult to imagine what limits to our Golems will remain.

In the early spring of 1998, students and faculty at Melbourne University in Australia must have been rather startled to see the following announcement tacked up on boards around campus.

TALK
"Artificial Brains and the 21st Century Species Dominance War"
Tuesday, 31 March 1998, 2:15 P.M.
Melbourne University Computer Science Dept.
Theatre 2, Basement, SEECS Building,
221 Bouverie Street, Carlton,
Victoria, Australia.
Dr. Hugo de Garis
Head of Brain Builder Group
ATR Labs, Kyoto, Japan.
www.hip.atr.co.jp/~degaris
> Dr. de Garis is the only scientific Davos (World Economic Forum) Fellow in Japan. One of his life goals is to make the planet aware that 21st century global politics will be dominated by the artillect issue. . . .
Summary of Talk
> Dr. Hugo de Garis, who was born and grew up in Australia, is head of the Brain Builder Group at the ATR Labs in Kyoto, Japan. His ambition in life is to make artificial brains using electronic neural

circuits that "grow and evolve" at electronic speeds in special hardware. 10,000s of these circuits can be evolved in a second each and later assembled into humanly defined artificial brain architectures. By the year 2000, Dr. de Garis hopes to have built the world's first artificial brain containing 10,000 circuits controlling the actions of a life-size kitten robot.

Dr. de Garis . . . is very worried about the long-term consequences of his brain-building work. He predicts warfare in the late 21st century between two human groups bitterly opposed to each other over the question of whether human beings should build "artillects" (artificial intellects) or not. 21st-century technologies will allow the construction of computers with 10^{40} components by using, single bit per atom, self-assembling, reversible, heatless, 3D, asteroid size, nano, quantum computers. Human beings have only 10^{10} neurons. Thus 21st-century computers have the potential to vastly surpass human intelligence levels.

In a nutshell, the question that will dominate the twenty-first century will be "Do we build gods or do we build our exterminators?" for the simple reason that, as de Garis sees it, the robots to come are likely to grow impatient with their second-rate creators. "We could never be sure these artillects . . . wouldn't decide that humanity is a pest and try to exterminate us, and they'd be so intelligent they could do it easily."[12]

Apart from the worrisome implications of quantum technology regarding our future, there are also implications as to our *nature*. If artificial quantum technology could generate such astounding means of handling *information,* and especially if that technology were to be in many respects *evolved* instead of designed, does it not open the possibility that nature has likewise learned how to employ quantum effects in our own evolution? Would we not thereby no longer need consider ourselves mere machines but rather creatures who, like the quantum Golems now on the horizon, embody a mysterious capacity to know as well as some of the freedom that quantum systems alone appear to possess?

The answer is "yes," but the brain is not quantum in any direct way—it is not a massive quantum computer of the sort we've just envisioned. The way that the human brain may plausibly embody and take advantage of quantum effects is so subtle as to be almost indetectable, but its implications are powerful. To understand how, we proceed deeper into the structure of the brain.

14

IN THE MATTER OF MIND

*The true and strong and sound mind is the mind that can
embrace equally great things and small.*
—Samuel Johnson, Boswell's *Life of Johnson,* Chapter VI (1778)

IS ANYONE THERE?

When young, I had a pretty fancy Gilbert microscope at home—at least it seemed fancy to me—and I loved peering at brine shrimp available by mail order during the winter and at the various single-celled denizens of a nearby pond when the weather warmed up. I was possessed of an insatiable curiosity about things, but I wouldn't go so far as to call it "a sense of wonder." There was no awe, just fun.

But then I had an experience that struck far more deeply. I don't recall exactly where or when it took place—probably early in high school—but the key images remain vivid (and were reinforced when later I studied such matters more carefully). Our class was watching a film of a cell dividing. The nuclear membrane dissolved and the chromosomal material at its core began to unravel. The film was in black and white, and grainy, and for a while the images were rather confused-seeming. It all looked like the vaguely ordered but largely random wigglings I had grown accustomed to seeing under the microscope, albeit now at an intracellular level.

Then the imagery changed. The nuclear material abruptly ordered itself, and two distinct symmetrical starbursts of gently curving lines appeared on either side. The starbursts guided and directed the nuclear material to double and pull swiftly apart in an elegant orchestration. (See figure 14–1.) It was as delicate, as powerful, and as cohesive as a corps de ballet at the apex of a performance. I was thunderstruck.

At the time, I didn't have much by way of a cognitive framework within which to put what I had just witnessed. My thoughts were something like this: Everything we studied about science suggested that intelligent life was a phenomenon that emerged only at the very highest levels of complexity. Indeed, in a manner of speaking, life itself seemed to be about intelligence—the more

FIGURE 14–1   Spindle formation during mitosis.

alive something was, so to speak, the more its behavior was intelligent. Yes, amoebas and paramecia and brine shrimp were alive too, but the less-aliveness of their lives was related to their simplemindedness. In fact, they were so simpleminded that one could almost feel how much closer they were to inanimate matter than a lizard or a dog or a person.

But watching cell division, especially the "mitotic spindles"—the symmetrical starbursts—and the orchestration that followed their appearance, I had the overwhelming impression that I was watching an extraordinary order and power—an intelligence—at a level beneath and within that of the simplest single-celled creature. Not only were single cells alive, there were things within them that were too.

Later on, in college biology courses and in medical school, I studied these processes more carefully, always hoping to understand something of this mysterious power. I saw many more films and videos of mitotic division, of ever higher resolution and clarity, and every time the hair on the back of my neck would stand up (and still does, even as I write). But these studies were always half disappointing. The halves of my reaction were as follows.

On one hand, molecular biology was advancing at a breakneck pace, and what we had learned about how living matter operates was mind-boggling and exciting in its detail and exactness. We had learned, for example, that the mitotic spindles that so fascinated me were composed of strawlike hollow tubes called "microtubules." Many effective cancer drugs (e.g., vincristine) work because they inhibit the development and functioning of microtubules, hence attack dividing cells, hence, cancer cells especially, since these spend so much time dividing wildly.

But, on the other hand, all the exciting advances grew out of the idea that all living processes were completely machinelike at the molecular level. The interaction of large numbers of machinelike processes were not planned but arbitrary—"random" was the term used. The beautiful choreography was an illusion—strictly in the eye of the beholder, who himself was a machine of randomly interacting parts driven willy-nilly for as yet unknown reasons to concoct illusions such as beauty and choreography.

These cold, impersonal premises were stunningly powerful in their ability to yield results. Yet I could never escape the awe whenever I watched cell division. No matter how I tried, the impression remained that something more than mechanism, something quite mysterious, lay at the heart of life. Perhaps it wasn't anything that we had ever thought of before—"vital essence," as the vitalists thought, or "Chi," as did the Chinese, or "prana" as in the Vedic tradition, or the "spirit" or "soul" or God—but it was definitely something other, and the impression it left of intelligence was unshakable.

NOT BY CHANCE

Part of the puzzle, I knew, had to do with the contrast—not really a contradiction, but a subtle tension—between the two premises of modern biology: machinelike behavior and randomness.

Machines, of course, are perfectly orderly; their essence is the *lack* of randomness. So where does the "randomness" come in? In truth, it doesn't. What is meant by "random" in science—what has been meant by random for centuries—is not random at all but pseudorandom. More generally, one uses the term "stochastic," meaning "subject to the laws of probability." But many things are stochastic without being truly random.

For example, according to mechanical determinism, every molecule in an air-filled balloon moves along a predetermined and absolutely fixed path. If one molecule hits another, only loosely do we call it a "chance" collision. What we really mean is that the two paths shared no common causes. Their collision was incidental in the sense that their paths were in no way related to one another. Nonetheless, because the paths of both were also wholly determined, neither could do anything but collide. Afterward, their paths are related in being affected by the collision, but still perfectly mechanical and determined. The interaction between them occurs only in an extremely restricted region of space—the point of collision—and during an extremely brief period of time—the moment of collision. Put differently, all interactions are local.* This is precisely what Einstein had in mind in describing reality as local; it is the premise behind the various equivalent parallel processing concepts we've discussed: purely local, nonorchestrated interactions can, it seems, give rise to global order with a seeming intelligence.

There is nothing random in the least about this quite machinelike picture. Only when we restrict our attention to just one of the two molecules does the other seem to appear out of the blue, hence "at random." That's only because we weren't paying enough attention. When we don't need to pay such close attention to the details, or can't, and if we are dealing with large enough numbers of identical things, we replace exact, mechanical laws with approximate, statistical ones. We pretend that the many identical things interact randomly, knowing full well that they don't.

---

* The definition of "local" can be extended to take into account interactions that arise at a distance due to forces that propagate. Gravity and electromagnetism are two such forces that require a so-called "field" extension of what we mean by local. The same principles hold, however. To accommodate these interactions we introduce a fuzzy boundary to every particle, whose "hardness" to each other changes as they approach one another. This makes the mathematics more complicated, but doesn't alter the principle: Just as one particle can't influence another without traveling to it at some speed never greater than that of light, the force field of a particle can't influence another without traveling to it, likewise at a speed never greater than that of light.

The mechanicality/randomness quality of modern scientific explanation—its reductive, lifeless "feel"—derives from the purely local character of interactions. If these interactions happen to perform some global computational "task," as do neural networks, spin glasses, cellular automata, and the like, then they look "intelligent." But there is no genuine "unity" among the parts, just something that looks like it to the observer concentrating on the whole. No conductor, no sheet music is needed, for synchronized waves to arise in a cellular automaton.

Einstein knew that a genuine global unity among nonlocal elements would have the opposite feel—that is, "religious" or "spooky"—evoking an impression in the observer of "something more" than the only-seeming intelligence of emergent, parallel computation. It's what we sense and admire in a great ballet or symphony. In these instances, there is something more than the mere interaction of individual dancers or musicians—there's the governing mind of the composer, conductor, choreographer: the idea, the creativity, the will: the consciousness. Of course, if there is no such thing as freedom, then there is no genuine mind, idea, creativity, will, or consciousness at work in these instances. That it seems so must therefore be a result of massive self-organization giving rise to the illusion of unity. In that case, the dance of the mitotic spindles is obviously another, vastly more elementary instance of self-organization.

From this we can draw two conclusions.

First, everything points to the likelihood that mitotic spindles in particular (but much else within the cell too) are in some fashion analogous to neural networks, or cellular automata, or spin glasses. That is, they must surely be self-organizing, hence, in some sense, computational entities within the cell. Should they turn out to be important enough, we might want to consider them the cell's interior nervous system (accepting the possibility that intelligence within the cell may arise in a different way and be less a specialty of just one system than in a whole organism).

Second, if there is more than illusion to the impression of "choreography" among the spindles—that is, of an intelligent choice among options, which is what choreography, indeed, any form of artistic expression implies—it will have to be due to some kind of quantum effects, because "there is no place for true randomness in deterministic classical dynamics," and without some source of randomness there are no options. Where is there such a source? Everywhere, ironically enough: because "fundamentally the universe is quantum mechanical."[1] It—all of the material universe in its quantum nature—is the source, the only source. It is everywhere yet seems to us nowhere, because apart from exceptional circumstances, that quantum nature is evident only at unimaginably small scales. If life somehow makes use of quantum effects to achieve freedom, it must have figured a way to amplify those effects upward across many, many scales.

Of course, some of the greatest of the quantum pioneers wondered aloud whether the difference between living and nonliving matter had something to do with quantum effects—the analogy is too compelling not to wonder this, even if in the end most scientists can find no evidence for such a connection. Their peers were in fact mostly doubtful, and biologists accused them of speculating about subjects of which they were in nearly complete ignorance, Nobel Prize status notwithstanding.

The notion that life was in some sense a quantum phenomenon arose in part from the analogies that may easily be made between the behavior of quantum particles and the behavior of living entities—especially people, as we've noted before. Poincaré, Heisenberg, Pauli, and many others often remarked on how like "free will" seemed the "decision" of quantum particles to show up "wherever they wanted." If quantum particles exhibited behavior that was truly caused by nothing, while living systems (people included) were mere machines, then might one not argue that atomic particles are rather the true beings possessed of free will, whereas we aren't?

To the extent they could, the quantum pioneers tried to back up their hunches with plausibility arguments. Bohr, for instance, remarked on how the ever-changing medium in which a wave is but a pattern is like the ever-changing matter in which life is a pattern too.[2] Louis de Broglie, who first came up with the idea of "matter waves" and won the Nobel Prize for it, insisted that "the structure of the material universe has something in common with the laws that govern the workings of the human mind."[3] But hard findings—even testable hypotheses—were hard to come by. In 1994 the outstanding British mathematical physicist Roger Penrose published a best-selling book, *Shadows of the Mind*, asserting that the brain is a massive quantum entity, not a mechanical, self-organizing automaton.[4] He, too, wanted a way out of the cage, as he had made clear in a worry five years earlier: "Perhaps we are doomed to be computers after all! Personally, I do not believe so, but further considerations are needed if we are to find our way out."[5]

But Penrose was not then looking for quantum effects within the cell. His later model, formulated in conjunction with Stuart Hameroff, a research anesthesiologist at the University of Arizona, required that some 20,000 neurons at a time exist for about one-fortieth of a second (in the quantum world, an enormous duration) in a single *coherent* quantum superposition—meaning that all the particles involved would behave as if they were one gigantic quantum entity, all their individual waveforms adding up so as to create one simple waveform that remained unchanged. (See appendix C for a more detailed explanation.)

It is inconceivable that the incidental jiggling of such an enormous amount of living matter, at 98.6 degrees F, would not overwhelm any quantum effects by causing almost all to cancel out. Penrose wasn't merely making an analogy between people and electrons, but still, the gap between known quantum effects and a living net of thousands of neurons is immense. For all the book's popularity—and in spite of Penrose's genuinely stellar reputation and contributions to physics and mathematics (only Einstein himself has contributed more to the Theory of General Relativity)—*Shadows of the Mind* was quickly assigned to the growing heap of pop physics and consciousness books that few scientists take seriously.

Indeed, it has been greeted with jibes and lampoons and howls of protest. In a not-so-sly caricature of the theory's 1960s Alice and the White Rabbit overtones, the unflinchingly materialistic, mechanistic philosophers Patricia Churchland and Rick Grush, at the University of California at San Diego, remarked that the Penrose-Hameroff idea "is no better supported than any one of a gazillion caterpillar-with-hookah hypotheses."[6] In their response, Hameroff and Penrose good-naturedly responded with their own cartoon depicting Churchland and Grush as ostriches, their heads deep in the sand.

The underlying clash of worldviews—mechanical determinism as true science, quantum mechanics as "religion"—as doped-out New Age all-is-one mush-type religion—is evident as a barely concealed subtext to many of Penrose's critics' barbs. Murray Gell-Mann, Nobel laureate theoretical physicist, longstanding colleague of Richard Feynman at Cal Tech and now at the Santa Fe Institute, put it this way:

> [W]hat characterizes his proposal, as far as I can tell, is the notion that consciousness is somehow connected with quantum gravity. . . . I can see absolutely no reason for . . . engaging in mystical speculation about quantum gravity. . . .
>
> Hardly anyone is left who thinks that special vital forces, apart from physics and chemistry, are needed to explain biology. Well, the idea that special physical processes are needed to explain self awareness will soon die out as well.[7]

With as unchippable an edge as ever, Marvin Minsky says simply of Penrose's idea that "Where there's smoke, a sharp reader could only conclude, there is smoke."*[8]

In the words of theoretical chemist Michael Kellman at the University of Oregon, one of a group of researchers who have performed hard analyses of the actual quantum dynamics of molecules, the problem is that:

> To be true, [the Penrose model] . . . would have to involve some completely marvelous structures about which we know nothing—just saying that neurons strung together might be coherent quantum systems is about as plausible as saying I can tunnel through Mt. Everest because I'm a quantum system.
>
> I'm not saying that *something* that would do what Penrose wants is impossible. Just that it's probably impossible right now to guess what it might turn out to be. One (many people, probably for centuries, maybe less) will have to think and do experiments to figure out how all of this might work.[9]

---

* "Where There's Smoke, There's Smoke" was the title of a 1979 article by the physicist John Wheeler on the 1969 decision by the board of the American Association for the Advancement of Science (to which Wheeler was a minority dissenter) to allow the *parapsychological* association recognition in AAAS. Wheeler considered parapsychology to be pseudoscience and still does. "The sharp reader" understands that Minsky is saying the same of the Penrose "science of consciousness." Of course, given Minsky's take years back on the potential value of neural networks, perhaps there's hope for the Penrose hypothesis after all.

John Hopfield, whose own work lies so close to the heart of everything we've been discussing, put it most bluntly: "contrary to the expectations of a long history of ill-prepared physicists approaching biology, there is absolutely no indication that quantum mechanics plays a significant role in biology."[10]

But Daniel Dennett, philosopher and director of the Center for Cognitive Studies at Tufts University, explains the heart of the matter: "Most biologists think that quantum effects all just cancel out in the brain. . . . *They don't amplify them; they don't hinge on them.*"[11]

Wrong as I think may be Penrose, *so is Dennett*. This should seem a contradiction, but it isn't. It all hangs on the distinction between "hinge" and "amplify" and what exactly could be meant by the latter. For that, we turn back to mitotic spindles and other parts of the cell much like them.

THE CELL'S CELL

Under the microscope, when it was first invented, a new truth became evident at a glance: The fundamental unit of life is the cell; of the nervous system, therefore, the nerve cell. This is the "neuron doctrine." Once computational neuroscience arose, the neuron was naturally understood as the most fundamental unit of *biological intelligence*. So long as we view the individual neuron as a relatively simple entity—it adds its inputs, transforms them according to a fixed rule, then distributes the result widely to its neighbors (perhaps also to physically distant friends)—the neuron *must* be the most fundamental unit. These functions do not require any sophistication or subtlety to implement; all the sophistication and subtlety emerge from the *interaction* of the parts—the whole.

But even a primitive single-celled organism exhibits an extraordinary computational complexity. Amoebas and paramecia drive toward goals, react, avoid—in their way, they even show signs of memory. Of what computationally capable parts, then, is the paramecium composed?

Like most single-celled organisms, the paramecium has an astounding array of differentiated parts. Our mental picture of a cell as a rounded square with a big black dot in the center accurately conveys the degree of structure within the cell about to the same extent as a smiley face conveys the nuance of emotion in the *Mona Lisa*. Nonetheless, one structure does stand out as a prime agent of the paramecium's intelligent-seeming behavior: its *cilia*. (See figure 14–2.) These are the fine hairs with which it is covered and that react to the environment. They are at once antennae and agents of motion, processing input from the organism's surround both to determine whether it should eat, fight, flee, or reproduce, and to generate the synchronous motions needed to engage in all four actions. The action and the intelligence are one. (Between paramecial thought and action, it seems, there is no reflection; not even a ten-count. For it, to feel an impulse is to act on it. But that's not so different from all but the most civilized human behavior.)

Cilia operate by generating synchronized, traveling, oscillatory waves—it should sound familiar. This complex synchrony arises from the purely local interactions of yet smaller parts. Every individual cilium, in other words, each of

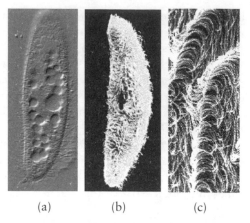

(a)   (b)   (c)

FIGURE 14–2   Paramecium. (a) Note internal organelles and external cilia.
(b) Scanning microscope shows cilia covering entire surface. (c) Detail of cilia showing
individual waves and coherent wave groups.

which demonstrates its own internal wave coordination, and a collection of cilia
that demonstrates group coordination, is engaging in massively parallel computation for all the reasons outlined before: self-organizing global order arising out
of local interactions of arrays of identical units; the establishment of stable oscillations within subgroups; and traveling waves implying temporal synchrony.

The cilia of the paramecium are *microtubules*, little different from the
hollow, strawlike structures of which mitotic spindles are made. Neurons, too,
are especially rich in microtubules where they sustain both signal propagation
and the self-organized construction of new neuron-to-neuron connections. At
three increasing levels of magnification, mitotic spindles, cilia, and other microtubular structures look like those in figure 14–3.

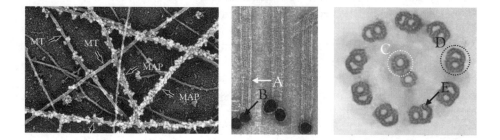

FIGURE 14–3   Microtubules forming "axonemal" structures (e.g., mitotic spindles, cilia in protozoa and dendritic and axonal signaling and transport mechanisms in
nerve cells). (left) "MT" marks individual microtubules; "MAP" marks microtubule
associated proteins. (middle) A: Enlarged microtubules forming segments; B: (-) charge
at end stains black. (right) C: The central axis of a cilium is composed of two microtubules. D: The circumference of a cilium is composed of 9 microtubules, each of
which in cross-section is composed of a circle and a half (large dotted circle). E: All microtubules are built up as a helix of a protein called tubulin (small solid circle).

Microtubules are hollow structures composed of many identical building blocks, proteins called "tubulin." Tubulin is a "dimer," a protein composed of two very similar but not absolutely identical subunits, referred to as α-tubulin and β-tubulin. A third variant, γ-tubulin, is less common, but plays a crucial role in the formation of tubulin branches and bends. In figure 14–4, from A to H, we see an ever more magnified visualization of the microtublule and its tubulin components.

The basic structural elements of the most primitive, most evolutionarily ancient single-celled organisms are likewise these same microtubules, suggesting even more strongly how fundamental they are to life. (See figure 14–5.)

A plausible argument may be made that the basic building block of cellular *intelligence* is the microtubule—so much so that damage to microtubules may result specifically in intelligence-eroding illnesses in us. For example, recent evidence suggests that within the neuron, the location of the primary pathology in Alzheimer's disease, is the microtubule: "While most investigators in the field believe strongly that amyloid deposition is at the core of [Alzheimer's] disease . . . a more coherent . . . schema [is] cytoskeletal [microtubular] abnormality. . . . [A]ll four identified genes interact . . . with the cytoskeleton . . . [and] abnormality of the latter leads to . . . loss of synapses and subsequent loss of neurons.[12]

Stuart Hameroff, a medical researcher at the University of Arizona, was one of the first (in the early 1980s) to propose that microtubules compute. It was an outrageous idea for its time. For decades microtubules were thought to be the *cytoskeleton*—the cellular skeleton, with its implication of a static, if flexible, but dumb scaffolding. As this was before self-organization of any sort was taken seriously, researchers automatically assumed that some yet-to-be-discovered central agency coordinated the movement of spindles and other related ("axonemic") structures—probably from within the nucleus. After all, bones don't think, brains do.

But it soon became clear that far from being static, microtubules are rapidly self-assembling and disassembling entities. They engage in complex, self-organizing, oscillatory, collective behavior that arises solely due to the nearest-neighbor interactions of adjacent identical constituents in an ordered array. From this perspective, Hameroff's claim that microtubules engage in massively parallel data processing began to appear less outrageous than obvious.

Just as the human fourth brain does not in every case require a master programmer to learn, adapt, and engage in highly sophisticated computation, neither does the "cytoskeleton." The tail of the human sperm (a microtubular structure very like the single or dual ciliary tails of certain protozoa, e.g., *Euglena*) wiggles because of propagating waves of conformational change that arise solely from nearest-neighbor interactions. These are very like the oscillatory waves among a collection of neurons (at a larger scale), or the oscillatory waves among a collection of chemicals (at a smaller scale), or the oscillatory waves of the business cycle (at the largest scale of which we are yet aware, these latter being so complex, however, as to be less rigidly regular).

Hameroff's insight was to see in microtubular interaction a key to intracellular self-organized intelligent processing. There are many other systems

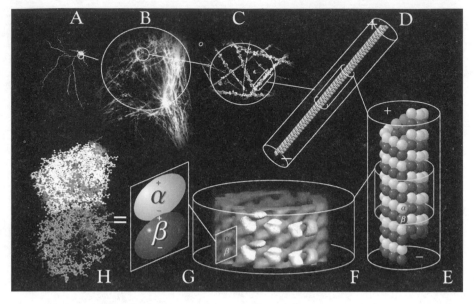

FIGURE 14–4   Microtubule (MT) network visualized by tubulin-specific stains (A, B, C edited and modified for clarity) or in model form.

(A) Stained MT networks ("neurites") in a neuron. At this scale, the letter "A" is ca. 0.15 mm across (= 1.5×10^{-4} m).

(B) Close-up of microtubular structure at the cell body, around the nucleus. The letter "B" is ca. 5 μ across (= 5×10^{-6} m).

(C) Individual MTs with MAPs ("MT-associated proteins"). The letter "C" is ca. 150 nm across(= 1.5×10^{-7} m).

(D) Model of a short section of MT showing helical structure. MAPs are not represented. The letter "D" is ca. 20 nm across (= 2×10^{-8} m).

(E) Detail of helical structure. Three strands start along a single diameter. α-tubulin and β-tubulin helices seem to alternate, but the stronger bond is the vertical one between one α and one β subunit. The triple helix means that there is a "seam" (shown in front) where α-subunits laterally adjoin β-subunits. The MT diameter is ca. 25 nm (= 2.5×10^{-8} m). The letter "E" is ca. 5 nm across (= 5×10^{-9} m).

(F) Electron micrograph of a section of a MT. The letter "F" is ca. 2 nm across (= 2×10^{-9} m). The light-colored extrusions are various MT associated proteins.

(G) Schematic of the tubulin molecule. Both subunits have a positively charged end and a negatively charged end. These align to leave a net electric dipole. MTs therefore also have a positively and negatively charged end, like a magnet, except the force is electrical. In between the nearby opposite charges attract and cancel. The letter "G" is ca. 0.5 nm across (= 5×10^{-10} m).

(H) Detailed stick-and-ball model of the tubulin protein. The letter "H" is ca. 0.25 nm across (= 2.5×10^{-10} m). (H) is magnified ca. 40,000 times more than is (A).

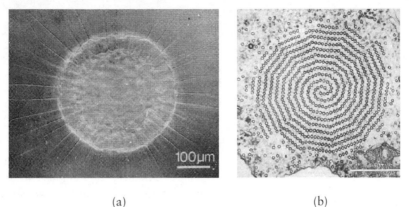

(a)   (b)

FIGURE 14-5   (a) *Actinosphaerium*. This present-day relative of the urchin is similar to those that appeared during the Cambrian "explosion," when life-forms mysteriously proliferated on Earth. It has ca. 300 radial axonemes. (b) Cross-section through one axoneme. Each axoneme is composed of about 720 microtubules (= 3 billion tubulin proteins) in a double-helical array.

within the cell that can be reconceptualized from a parallel information-processing point of view: the DNA-RNA-protein cycle most evidently, but biochemical cycles themselves as well. Be that as it may, the distinctive structure of microtubules, as we will see, makes them a particularly appealing candidate as the next link after (below) neurons in a chain that connects the human macrocosm to the microcosms out of which it is built.

NETWORKS OF MICROTUBULES (MTs); MTs AS NETWORKS

The more one studies MTs, the more amazing they reveal themselves to be. They "are ubiquitous . . . and they perform an astounding array of functions."[13] These functions include:

■   The intracellular transport of organelles. The MT network is like a rail system along which specialized proteins are cargo-hauling locomotives. But this organic railroad is inherently intelligent. It requires no dispatchers or engineers; it assembles, disassembles, and routes itself on an as-needed basis.

■   Ciliary and flagellar motion. Movements of these structures are undulatory and synchronized, that is, traveling wave patterns with computational analogs at lower chemical scales (chemical oscillations) and at higher neuronal scales (neural networks).

■   The orchestration of chromosomal division during cell replication. The rhythmic "dance of the chromosomes" is an example of synchronized, oscillatory movement within the cell, with the same lower- and higher-scale computational analogs as give rise to cell motion.

■ Changing signals from one medium to another. This is a function that MTs perform in all cells. But a highly specialized kind of microtubule is required in neurons for them to establish new connections.[14] Indeed, "The cytoskeleton is a natural candidate for the task of carrying directed signals because it spans the entire cell and geometrically speaking it is a branched one-dimensional structure."[15]

■ Communication between the nucleus and the cell interior. Transduced signals in the MT network consist in large part (though not entirely) of self-organized changes in the shape of the MTs and their layout. Signaling and shapeshifting are one and the same. The ends of a microtubular network are physically attached to parts of the cell (e.g., organelles, the nucleus, the membrane) such that the parallel processing, signaling, and transduction of physical force form one seamlessly integrated function:

> a team of cell biologists . . . has demonstrated that mammalian cells are densely "hard-wired" with force-carrying connections that reach all the way from the membrane through the cytoskeleton to the genome.
>   . . . The experiment, by Andrew Maniotis, Donald Ingber, and their collaborators,[16] is being hailed as a triumph. . . ."It puts the cytoskeleton in a new light: as a mechanism for signal transduction rather than just as a supporting mechanism." . . . The mechanical communications system . . . extends through the nucleus [to] individual chromosomes.[17]

Recent techniques allow the visualization of MTs in living cells, and these show that conformational changes in the cytoskeleton actively shape and reshape its internal configuration to create mobility. Similarly, the physical layout of growing neuronal dendrites is guided by the self-organizing physical patterns of its MTs and associated proteins and filaments.[18] As one Harvard researcher put it, "[A] coherent picture is finally beginning to drop out . . . a 'thinking' signal transduction system . . . integrates multifaceted and sometimes conflicting information to come up with the appropriate biological response."[19]

In this way, the neuron's internal microtubular network plays a key role in the wiring of the macroscopic neural network of the brain. The "signaling" along microtubule networks consists, inter alia, of patterns of mechanical tension—and corresponding changes in shape—transduced and transmitted through the entire cell system. This makes perfect sense, of course, because if, in general, the hardware and software of a densely connected network are one and the same, we would expect basic mechanical relationships to play an ever-increasing role in computation at smaller and smaller scales.

For example, chemical kinetics and therefore the newly emerging field of molecular computation are based largely on the interplay of electrical forces between molecules. Input stimuli to such a computer are necessarily going to cause relatively large-scale changes in the position, motion, conformation, and chemical features of the "stuff" that makes up the computer.[20]

Because at small scales, physical motion of the parts of the thing that's doing the computing is a major form taken by the computation; and because

MTs are at such a scale, they are now under study as a means of creating nano-level, self-organizing robotic machines.[21] These are robots for which the computational medium and the structure whose motion is directed by the computation are one and the same. Their role within the cell appears precisely the same.

THE LANGTON CA CONNECTION

Since 1982 Hameroff had been looking at MTs as potential cellular automata.[22] A few years later, Christopher Langton (who had designed the first self-replicating cellular automaton, and upon whose basic design Hugo de Garis's self-evolving CA brain is based) edited a book on the exploding new field of "artificial life."[23] Langton was especially interested in the ways that aggregates of *molecules* could compute so as to self-assemble and self-organize into living matter. He had recently published an influential article on the subject in a journal directed primarily to physicists.[24] Included in the book was a long and detailed chapter by Hameroff and two colleagues entitled "Molecular Automata in Microtubules: Basic Computational Logic of the Living State." As the title implied, the authors hypothesized that, given the cell as the basic unit of life, its computational capacity was embodied in a distributed network of MTs functioning as a cellular automaton.

The underlying structure of MTs does indeed bear a striking resemblance to cellular automata. Each tubule is constructed out of thousands of tubulin dimers arranged in a "rolled-up" two-dimensional grid, as in figure 14–6.

Each tubulin molecule could then be thought of as an automaton cell. Because of the geometry of the dimers, the resulting grid formed hexagonal neighborhoods (six nearest neighbors). The positive and negative electrostatic charges on the dimers caused them to align; the shifting of an electron between the α and β subunits was responsible for a conformational change in the dimer as a whole. The change of one dimer from one form to another exerted an in-

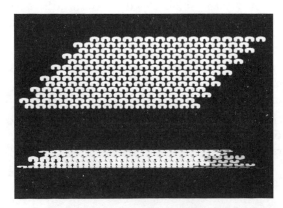

FIGURE 14–6   Schematic representations of a microtubule. (bottom): Rolled-up dimers form a tube. (top): Unrolled, the dimers form a sheet that is 13 dimers wide.

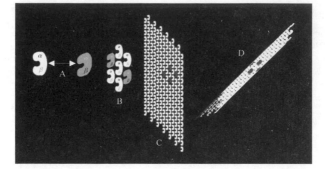

FIGURE 14–7   A microtubule as a cellular automaton. (A) The conformation of
the tubulin dimer changes depending on where a free electron resides (dark lettering).
The α-configuration is to the left, the β-configuration to the right of the double arrow.
(B) Six tubulin dimers forming a cellular neighborhood. (C) A pattern formed by four
tubulin dimers in the β configuration (free electron in the β subunit). (D) Rolled-up
grid of tubulin dimers forms an MT.

fluence on its neighbors. Thus, the MT had the basic elements of a two-state
cellular automaton. (See figure 14–7.)

    For the most part, cellular automata as then understood required a
"clocking mechanism"—something that triggered the simultaneous updating
of every cell based on the state of its neighbors. This means that there would
have to be some means for carrying a regular signal—like a vibration with a
fixed beat. But under a microscope, a network of MTs looks almost completely
disordered. It's hard to imagine how so chaotic a jumble could sustain anything like regular vibrations.

    However, a careful study of how MTs are linked one to the other reveals
a surprisingly harmonious order. MAPs (MT associated proteins) and inter-
linking filaments are positioned at quite (but not perfectly) regular intervals.
Two naturally occurring patterns of proteins are located where stationary
nodes ("frets") would have to be to sustain coherent standing sound vibrations.[25] (See figure 14–8a and b.)

    Furthermore, at least in single-celled organisms, the network as a whole
has a highly symmetric structure that would facilitate sustained vibrations. (See
figure 14–8c.)

    At the time, Hameroff's hypothesis of MT computation and coherent vi-
bration seemed far-fetched. But, as referenced earlier, we now know that MT
networks do indeed transmit and transduce fluctuating forces, maintaining almost instantaneously a constantly shifting, dynamic equilibrium across the entire network. Since signaling and shape-change are the same in MTs, it would
make sense for coherent vibrations transmitted along the MT to serve a related
function: As a vibratory wave passed a region, it would energize the tubulins
there to change their conformation. Rhythmic waves might then constitute a
"clocking" mechanism that timed and jarred the "updates" for the tubulin array as a cellular automaton. Assuming certain "rules" to govern nearest-neighbor interactions, Hameroff and his colleagues showed that stable propagating
structures could exist, of the sort shown in figures 14–9 and 14–10.

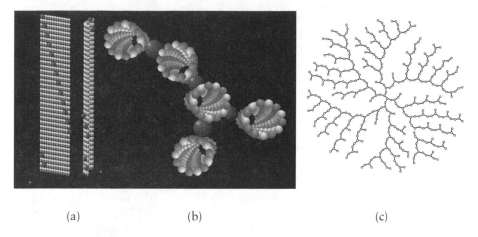

(a)   (b)   (c)

FIGURE 14-8   MAPs (microtubule associated proteins) forming attachments between individual tubules. (a) Unfolded MT section showing attached MAPs. (b) End-
on view of segments of five MTs attached by four protein bridges. (c) MT network in
a *Radiolaria*. Circles are MTs; lines are linking proteins.

FIGURE 14-9   Microtubule cellular automaton models. Light cells are tubulins
with an electron in the α subunit pocket. Dark cells are tubulins with an electron in the
β subunit pocket. Making certain assumptions, these are some of the stable patterns
predicted by Hameroff, et al. D1–D3: three states of a traveling wave. A–C: gliders.

FIGURE 14-10   Hypothetical traveling tubulin state wave in a microtubule.

There is a problem with this model, however: The rules governing inter-actions between tubulins, while not biologically impossible, are not those found in nature—they were hand-designed to ensure cellular automaton–like behavior.[26]

And yet Hameroff's insight—that MTs function as parallel computers in *some* way—is fundamental and unavoidable; and for that fact there *is* evidence. The question remains, how?

THE DEVILISH DETAILS

Between the plausible and the actual there is a huge gulf: Most of what is possible just ain't necessarily so. Jack Tuszynski at the University of Alberta had a solid background in the physics of complex systems and had established himself as head of a significant laboratory group. Like Hameroff, he had already focused a great deal of attention on MTs, understanding that they displayed all the necessary features of an intracellular intelligence.

Hameroff's model of MTs as cellular automata had required a large number of hypothetical assumptions: Yes, the MT structure *looks* a lot like a cellular automaton and behaves in some ways like one, but isn't. Tuszynski had a more physiological—and therefore more *natural*—take on the problem. It was the same process of discovery that led to the conviction that neural networks could compute but that the brain's neural network wasn't of the back-prop sort. If you want to understand nature, and learn from her, your models must incorporate the parameters within which she actually works—in as much detail as you can muster. The problem (expressed explicitly in a later paper by Brown and Tuszynski) was that the "Hameroff-Rasmussen-Mannson scheme is a zero-temperature model in which signal propagation is an artifact of the model's design."[27] For it to work—for the MT model to compute—you needed to assume that the MTs were living in an environment at 273 degrees C below zero—hardly compatible with biological systems.

Furthermore, it turns out that the *precise* effect of one tubulin on another, as well the *exact* geometry of their positioning vis-à-vis one another, has a dramatic effect on the information-processing potential of an MT—just as does the precise update rule and exact neighborhood in a celullar automaton. Like Hopfield, Tuszynski's familiarity with both physiology and physics was a great help. An important clue was the general fact that *if a system should turn out to behave like a spin glass*, "[I]nherent in the dynamics of the physical system of interacting spins are the properties that enable it to represent and compute,"[28] in the words of Terence Sejnowski, who developed both NetTalk and the application of annealing to Hopfield nets.

Spin systems, of course, are not hand-designed, and MTs turn out to be much more like spin glasses than anything else, because, like a spin glass, and like those cellular automata capable of synchronization, they utilize competition and cooperation at the same time—that is, "frustration."

In 1993 Tuszynski and a group of colleagues sketched a new and far more detailed portrait of what might be going on in MTs, a portrait that incorporates all the then-known details of real MTs.[29] What falls out of the Tuszynski model is quite astonishing: Because physiological temperature is what it is (*not* absolute zero!); because of the precise parameters involved in the relations among tubulins in a neighborhood; and because these parameters favor an odd-numbered (13) unit circumference with skewed hexagons forming a helical tube, the unique MTs of the human nervous system can function simultaneously as *electromechanical signal transmitters along their length and as a spin glass type parallel computer helically, about their circumference.* If you think of it as a cellular automaton, then it's a cellular automaton that is skewed slightly, then rolled up to form a tube with a longitudinal seam. The seam causes circumferential frustration; along its length it allows propagation. In the end, whatever outputs it computes in response to inputs, it transmits. As a "wire," it's a *smart wire* of the sort that nanotech research is attempting to develop. It *processes the signals adaptively as it transmits them.* (The details of how this works may be found in appendix B.)

Furthermore, the adapting pattern of activity of a neuron as it participates in a living neural net alters the signal processing happening along its own internal microtubule network. The processing characteristics of all individual neurons in the net are likewise affected by changes in their microtubule network. In short, the neural network of the brain is coupled to the internal neuronal microtubule network and vice versa. The microtubule network therefore adapts to the demands of the brain as a whole, with neuronal plasticity as the intermediary.

What's the advantage of a smart wire over a dumb one? As we have seen repeatedly, smart structures learn and adapt. If MTs are in fact a key part of the development of neural structures, it would be suitable that they be able to adapt to the pressures that are transmitted throughout the network as a whole. Instead of having a minimum of adaptive machinery, allowing learning and intelligence to develop only at the level of the network of neurons, each neuron would itself behave like an adaptive network—just as each person in a social network is him- or herself a complex network of processing elements.

BUILDING AN MT BRAIN

Michael Conrad is head of the Biocomputing Group at Wayne State University, which models complex biological systems at various levels of organization. Their mission is to develop molecular computers employing the same principles as living computation. Like Hugo de Garis's group in Japan, one of Conrad's projects is the development of a self-evolving brain. Unlike the de Garis brain, which depends on a strictly cellular automaton model (that evolves a neural network), the Conrad brain employs methods that mimic both interneuronal neural networks per se and *intra*neuronal information processing, specifically that apparently employed by MTs.

In so doing, the Conrad team is specifically addressing the ways in which the pressure to adapt (the boundary conditions of a cellular automaton or a Hopfield net that constrain the energy landscape; the supervised learning of a back-propagation network; the conditions that determine which of HER jelly-beans are eaten, and when)—that is, evolutionary pressures—coordinates adaptation *within* neurons as well as *between* them.

Just as Rosenberg modeled his Perceptron on what was known about the retina, Conrad's group incorporates the equivalent of MT assembly and disassembly into their computational processes, as well as the biological relation between electrical signals and physical conformation. The result is a computational system that, even in rudimentary form, successfully classifies patterns, just as do Hopfield networks or self-organizing maps, and that responds very much like a growing neural network whose efficiency derives from the specifically *oscillatory* nature of its elements—the conformational "flip" of tubulin dimers that makes microtubules so like spin glasses:

> The MT network . . . supports a combination of growth dynamics and signal transmission . . . form[ing] an intrinsic adaptive loop. If the transformation of input signals to output is unsatisfactory the growth dynamics are altered, thereby altering the signal processing. The sequence of structural changes and alteration in processing continues until the system converges. . . .
>
> [A] . . . model . . . [of] MT adaptation . . . comprises . . . growth dynamics, signal propagation and . . . adaptive self-stabilization. . . .
>
> Signal propagation uses the structure generated by the growth dynamics. Individual MTs are modeled as strings of discrete oscillators with neighbor-neighbor interactions. Signals propagate both as longitudinal waves along MTs and between microtubules through MAP mediated connections. . . .
>
> Adaptive self-stabilization is somewhat like simulated annealing, but with error feedback. . . . Classification of input signals has been . . . [accomplished]. The model is crude in comparison to the natural system; nevertheless it demonstrates how natural it is for the microtubule network to exhibit *adaptive signal processing*.[30]

It would seem, however, that even having dropped down to the scale of structures that are but a few dozen billionths of an inch around, there is still no evidence of quantum effects playing an explicit role and no need therefore to appeal to quantum weirdness to explain such things as free will. True, the conformational changes in tubulin depend on electrons tunneling back and forth between α- and β-dimers. Does this imply that they therefore can do what "quantum computers" can? Not exactly—not what Penrose and others have hoped. But they can do something quite unique that does indeed bring the quantum world into play and eventually up into our own world. We now take the last jump down in scale into the brain—to that level where quantum effects are both crucial and ubiquitous—to within the elements that compose the microtubules of its neurons.

15

THE PROTEAN COMPUTER

Molecular biology is mankind's attempt to figure out how God engineered his greatest invention—life. As with all great inventions, details are top secret; however, even top secrets may become known. I find it a great privilege to live in a time when God allows us to gain some insight into His construction plans, only a short step away from giving us the power to control life processes genetically.

I hope it will be to the benefit of mankind, and not to its destruction.

—Arnold Neumaier, Institut für Mathematik, Universität Wien (1996)

Bob is an enlightened nerd. He has grokked to the fact that the universe, himself included, is a giant machine, and that free will is an illusion—along with God, gods, spirit, whatever. This bugs him, because he's thinking of getting married and doesn't want his bride to have been someone forcordained for all time. So he's figured out a way around the mechanicality. Bob has set up a double-slit experiment and has turned down the electron gun until, on average, only one particle a week is emitted. He leaves both slits open, and waits.

So far, everything he's done he *feels* to be freely chosen. But he also knows that it hasn't been. Since the beginning of time, it has been inevitable that Bob would exist, that he would set up an experiment in just this way, at just this time—even that he would laugh about the fact that he thinks of himself as having chosen to do what he's doing while knowing that he hasn't. He even knows that, because of quantum effects, there is some probability that he *might* have done something significantly different. But he also understands that because of all the averaging-away that goes on, that's a probability so remote as to be beyond consideration.

So Bob has put quantum probability to work for him. He's rigged the screen beyond the double slits. Each side, to the left and to the right of the midline, is connected to a computer and in both computers there sits the identical e-mail message ready to be sent: "I love you. Will you marry me?" One, however, is addressed to Alice, the other to Alison. If the next detected photon hits the left side of the screen, the message to Alice will be sent, the message to

Alison erased. And if the next detected photon appears to the right, the message to Alison will be sent, the one to Alice erased.

Instead of flipping a coin or asking a fortune-teller or "deciding for himself" (an illusion!)—all of which yield iron-bound predetermined results—Bob, for better or for worse, will be the first man in human history to *freely* ask someone to marry him, *and he will thereby create*, as has never before happened, *a different future* for himself and all future generations, based on the only genuine example of a random, nondetermined phenomenon in all the universe. Even though it was predetermined that he would do this, Bob will have eliminated the averaging-away of quantum effects and forced a single such event to choose one of two tremendously different large-scale futures. Such are the consequences of a *quantum amplifier*. Not until human beings came into existence, and not until they developed quantum theory, and tested it, could such an amplification been known to be possible at all.

Of course, to make the marriage entirely without cause, whoever gets his message should repeat the experiment with her own two e-mails: "Yes" and "No." (If the latter comes up, Bob may well consider *himself* to have been averaged away. But no matter.) Whatever the response, Bob's initial act of quantum wooing, requited or not, will yield an outcome determined by nothing.

QUANTUM BIOLOGY

Ever since quantum mechanics was first appealed to (by its inventors) as having some mysterious connection to life in general and to human freedom in particular, skeptics have derided such claims, as we've seen. Quantum events have nothing whatsoever to do with biology at even a microscopic scale, they argue, let alone with actions on the human scale. But Bob's approach to romance allows us to see that the skeptics are flatly wrong and that the evidence lies right under our noses: *Human beings have demonstrably been quantum amplifiers ever since they first concocted experiments*—like the double slit—*that generated dichotomous quantum outcomes at everyday scales.*

Now, a 100 percent mechanical system cannot under any circumstances generate an indeterminate outcome. Any true freedom of action *must* exist somewhere within the system—in and as an aspect of the configuration of its material elements. The only known source of perfect freedom of action resides in the quantum nature of matter. Should human beings be shown to have any capacity whatsoever to act other than in accord with rigid, mechanical determination, that capacity would therefore either have to derive from the quantum nature of matter, amplified, or arise from some other, utterly mysterious source. But, in fact, the only definite instance where human beings do generate actions that are irrefutably not predetermined are in situations such as Bob's—that is, when we amplify quantum events upward into our own scale of existence.

Here's what we conclude:

- The fact that human beings indeed have a capacity to elude mechanical determinism has at long last been demonstrated—even though that's not why quantum physics was developed or how it's been looked at since.

■ The only occasions where such freedom is certain are those when we know we are acting as quantum amplifiers. Therefore, the only certain human freedom arises from quantum events, amplified.

Though they may not seem to at first glance, these conclusions bear directly on biology and neuroscience and their possible relation to quantum effects. We can start with an obvious objection to the above, namely, that "deliberate" human constructions of quantum experiments and devices are "artificial." But stand back for a moment and reconsider. It's all a matter of perspective and of scale. Human beings are biological systems. At some point in their evolution (in the blink of an eye, by cosmic scales of time) their capacity to amplify quantum events became evident. Hence *these* biological systems (at least) clearly have had such a potential embedded in their material nature all along and must have evolved in such a way that the potential could be realized. It makes perfect sense, therefore, to look for such quantum events within the specific material configurations of matter of which these systems are constructed.

In light of the scientific consensus that man's makeup is material and the forms assumed by its material components essentially no different from the matter and makeup of other terrestrial life; in light of the consensus that man's *brain* in particular is qualitatively no different from the brains of many other "lower" animals; in light of the consensus that man's brain evolved from comparable "lower" material structures; and in light of the fact that, specifically, the evolved, material human *brain* dreamed up and designed the only known quantum amplifiers in the universe—it also makes sense to look to the brain and to its substructures as the living system that should most effectively have evolved to take advantage of quantum effects and to amplify them upward.

There remains one a priori objection to these ideas: Perhaps human beings really are somehow different; our experiments truly "artificial" and therefore no index of what biological systems can evolve. If so, then we must attribute to human beings some mysterious additional source of freedom of action that is not only not biological, it goes beyond even that which science can detect and confirm. But if we do that, then we assert, without any evidence whatsoever, that man is mysteriously free—just to avoid the possibility of quantum effects playing a role in living matter.

In brief: Unless quantum theory should be proven wrong, the existence of confirmatory quantum experiments proves the point that life has evolved nervous systems capable of amplifying internal quantum effects upward; and also the claim that (at least some) nervous systems can thereby evade mechanical determinism. To deny this does not restore mechanical determinism, as most critics think. It rather restores belief and rejects out of hand contrary scientific evidence, however obvious.

There is an eerie consequence to Bob's experiment. Our brains are mechanical, goes the determinist's argument. But we can hitch our fate to quantum indeterminism and commit actions that are not predetermined. But what if we go one step further and build artificial brains that, unlike our own, are based *directly* on quantum events? Might not such brains be not only smarter than ours, but *superior*? Perhaps, as it is claimed, it *was* God's intent to create creatures in his own image; creatures who, like himself, could throw the quantum dice as a way of life. It's just that we were mistaken in thinking these creatures are ourselves when, in fact, they are our progeny. Which still leaves us, the biologicals, once removed from freedom, right?

Not quite.

QUANTUM COMPUTERS OF THE SECOND KIND

Imagine a new kind of Rubik's Cube. Its twenty-seven component-cubes are actually magnets of different strengths. Your task is to arrange them so the net repulsive force among all the magnets (the energy) is the lowest possible. And you must do so by feel. If this sounds difficult, consider that nature solves a far more complex cube in seconds, uncountable times per day, to keep you alive: *the folding of proteins into their proper conformation.*

The importance of proteins cannot be exaggerated. If DNA is the blueprint of life and RNA the architect, proteins are at once preassembled construction materials, the factories that do the preassembling, the general contractor, the foremen, the subcontractors, the craftsmen, and the craftsmen's tools. They deliver raw materials to the factories and crews to the construction site. They need no as-builts, elevations, or building permits. They need no project schedulers, dispatchers, or quartermasters. Everything is built according to code, on time, without supervision. Their budgets undercut everyone. The amount of waste is staggeringly small. They protect and care for the architect and see to it that he reproduces himself and therefore their own replacements. They are the hard-wired security system and the instant response team. And all their functions are dependent on a precise shape and changes to it. For example, to defend an organism, they recognize and disable foreign antigens by changing shape and locking them in a death grip, then remember for a lifetime the shape of the invaders. To produce signals, they must recognize and bind to chemical messengers, send a signal, then let the messenger go. In this way, they prime the system for the next transmission and recycle the messenger to be used again.

Shape is critical to a protein's component parts, as well. Each is a string of smaller molecules called amino acids. Each amino acid has a unique shape and a unique distribution of charges. For a protein to work properly, the floppy string of its hundred or so amino acids must fold up on itself in just the right way—which most evidence suggests is at, or extremely near, their absolute energy minimum (like our magnetic Rubik's Cube). Why is this problem so difficult? Because DNA encodes the sequence of amino acids that compose a protein but doesn't contain a blueprint for the correct conformation or the correct sequence of folds. And a modest-size protein will have *at least* 400 trillion different shapes within a hair of the correct one.

To get a sense for the complexity of this problem, let's take a short tour of amino acids and their relation to neural networks and spin glasses, at the end of which we'll be better able to understand how a mindless chemical can solve such a huge calculation.

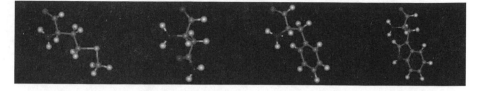

FIGURE 15–1   Four amino acids (from left to right): tryptophan, methionine, asparagine, phenylalanine. The balls are atoms of a small number of elements (hydrogen, carbon, oxygen, nitrogen, and occasionally sulfur); the sticks represent chemical bonds formed by the electrons of the atoms.

In human beings, all proteins are built up out of twenty-two different amino acids. Figure 15–1 shows four.

To construct a protein, the cell reads off the amino acid sequence encoded in a section of DNA, grabs the amino acids from the available pool (mostly from what we eat), and strings them like beads. One amino acid attaches to the other by forming a specialized bridge (a "peptide"), so the protein is truly one molecule. When incorporated into a protein, with a portion contributing to a peptide bridge, an amino acid is referred to as an "amino acid residue," or just "residue." In figure 15–2, from left to right, are a protein fragment composed of four amino acid residues (plus intervening peptide bridges) and two complete proteins, one small, the other medium size, composed of a great many residues and peptides.

Let's compare one of the simplest amino acids, alanine, to ammonia. Recall that ammonia has a continuum of shapes with two favored (lowest-energy) ones. All of its states can be represented along a single line that represents the position of its one nitrogen atom along one axis. Ammonia might vibrate a bit, but as it is built entirely of triangles, it is extremely stable. It can assume no configuration other than a pyramid of varying height. In short, it has very few "degrees of freedom" and, practically speaking, just one: where along the axis is its nitrogen. By contrast, figure 15–3 shows alanine, with just *some* of the internal motions available to its constituent atoms.

Taken together, these make alanine a very complex "clock." With this in mind, you can begin to understand the number of motions available to the sim-

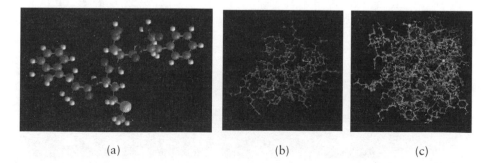

(a)   (b)   (c)

FIGURE 15–2   Amino acid strings. (a) 4 amino acids; (b) a small protein; (c) a midsized protein.

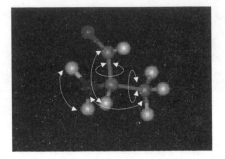

FIGURE 15-3   Some of many "degrees of freedom" available to alanine.

plest of proteins. Proteins therefore assume an astronomical number of conformational states with a very large number of relative energy minima. In an "energy landscape," every degree of freedom of every atom and bond is its own dimension. A Rubik's Cube—even our magnetic one—has a mere nine degrees of freedom. Nonetheless, within seconds, proteins find the correct sequence of folds to achieve their minimum energy state, the most extraordinarily complex minimization problem imaginable. But how? Not simply via trial and error—jiggling at random to try various conformations. That would take more time than the universe has existed. In spite of decades of intensive research at major universities all around the world and the application of massive supercomputing power, it is still completely impossible for us to discover from scratch the minimum energy configuration for even a short chain of amino acids. "[The] 'protein folding problem' has been singularly proof to the traditional methods of molecular biology and theoretical chemistry and physics."[1]

ORGANIC ORIGAMI

The sequence of amino acid residues is the protein's "primary structure." The correct three-dimensional shape is its "tertiary structure." DNA encodes the primary structure for every protein. About their final shape it knows nothing. Each peptide link can twist and turn almost at will, yet with blazing speed proteins somehow end up properly folded.

Between the simple sequence and the final shape, there is a "secondary" level of structure. Subsequences of residues form regions of *ropes*, *sheets*, or *helices*, as in figure 15–4.

FIGURE 15-4   Protein structure. (left) three secondary structure types. (middle) 1^0 & 3^0 structure. (right)1^0 & 2^0 structure. Fuse images to view in stereo.

FIGURE 15-5   Simulation of protein folding. The folding sequence shown above is hypothetical.

Since proteins are synthesized without folding instructions, they must proceed through a variety of possible folded states (like Rubik's Cube) to find the lowest-energy one. (See figure 15–5.)

Keep in mind, too, that the lowest-energy configuration is also the most highly functional configuration. In other words, many sections of ropes, sheets, and helices that are scattered far apart in the straight-line, primary structure get brought close together in the final, tertiary structure, to form pockets and extrusions of coordinated shape and charge—all without a blueprint other than for the linear sequence. The protein albumin that constitutes egg white is folded just so by nature. When cooked, the attractions and repulsions between residues and regions that constitute the protein's minimal-energy shape are overcome by thermal agitation, and it refolds, or "denatures." The clear, viscous liquid congeals to a semi-solid. One small piece of evidence for the subtlety of form in albumin is its amazingly unique viscosity and transparency. Another is the fact that being so clear (in spite of being so complex) and being composed of terrestrial amino acids, all of which have a left-handed form, albumin polarizes light. Spread a thin sheet of albumin between two pieces of glass and you have a rotary polarization filter.

Because the folding pattern of natural albumin is so complex and precise, reheating and cooling the egg back to room temperature won't reverse the process: Cooked egg-white proteins have settled into a far-higher-than-minimum-energy state that is nonetheless a relatively deep well—a "volcano crater." So how do individual proteins manage to find the correct conformation and avoid (innumerable) traps?

THROUGH THE LOOKING GLASS . . . AGAIN

Only after the development of neural networks, cellular automata, and the neural-network/spin glass mapping did the crucial clue emerge: Protein folding has many features of a spin glass–type minimization problem.[2] Proteins *anneal* to their minimum energy.[3] But, here's the kicker in this: *Their speed and efficiency in traversing the intermediate states that are part of any annealing process simply lie outside the range of any known classical process.*[4]

There is another puzzle, too. Many important functions require that a protein rapidly change between two or more equally or almost equally minimal energy conformations—like ammonia flipping inside out to oscillate between its two identical energy minima, only involving a fantastic number of axes along which the protein must "flip," not just one. This is true for every enzymatic reaction in the body, as also for the tubulin conformational changes that underlie microtubule function.

Furthermore, to leave an energy minimum requires energy (imagine reordering all your books; or climbing up and out of one alpine valley and down into another), yet proteins can make these changes using as little energy as is released from the breaking of just a single chemical bond. (Imagine putting just the one book you own by Abe Aaron in its proper location and having all the rest arrange themselves alphabetically.) This means that the energy is allocated somehow with exquisite precision and in such a way as to trigger a sequence of conformational changes each of which is tuned just so as to provide the needed energy for the next. Tubulin, in particular, changes its conformation back and forth between its two minima in response to the motion of but a single electron that directs the absorbed energy of a passing mechanical wave with a precision that puts Michael Tilson Thomas to shame. Attempts to model this efficiency using wholly classical means have simply never worked.[5]

The answer to the puzzle was slow in coming: *Proteins take direct advantage of quantum effects to do things that would otherwise be utterly impossible*. In particular, they take advantage of *tunneling*.

FANTASTIC AND FUZZY

The atom was once modeled on the solar system, the electron-as-planet revolving around the nucleus-as-sun. But remember: Electrons don't really have discrete orbits (i.e., trajectories). They act like delocalized waves (as we've seen) that occupy "orbitals"—fuzzy configurations of probability amplitudes, as in double-slit experiments. For example, a hydrogen atom could be drawn as in figure 15–6a.

Given a certain overall energy, the probability amplitude for the electron at a location is proportional to the density of the stippling there. The upper oblate sphere has positive values; the lower one, negative. The actual probability of finding the electron at some point is equal to the square of the probability amplitude at that point. (Both probabilities are therefore positive.) This does not mean that the electron *is* there before we find it: It is in a superposition of all possible locations; that is, it's delocalized.

Furthermore, when atoms combine to form molecules, their overlapping probability amplitudes are added *before* squaring. This means that separately, two atoms could have regions where there is a large chance of finding an electron, yet when the regions combine, there is *not* a greater chance of finding one there; there is in fact less chance—or none at all. A positive probability amplitude for the region near one atom can overlap with an equal but negative probability amplitude for the other, adding up to zero. This is the same phenomenon as in wave interference and can occur only in quantum systems. Depending on the energy of the system as a whole, delocalization and interfer-

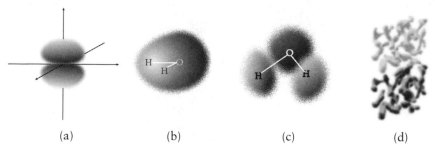

(a)   (b)   (c)   (d)

FIGURE 15-6   (a) $2P_z$ hydrogen orbital. (b) and (c) Two possible orbitals for water. Note the different extent of delocalization. (d) One of many orbital configurations for tubulin.

ence generate different and unexpected shapes for an electron "cloud," as in figure 15–6b–d.

Suppose that an electron starts out belonging more to atom A than to B—relatively localized, we might say. It then enters into a higher-energy superposition that is more or less evenly distributed (delocalized) between both A and B. Finally, when it yields back the energy, it can relax into a state where it is localized more to B—if this happens to be more energetically favorable. Even though nothing has "moved," and there are no "trajectories" involved, we say that atom A is the "electron donor," atom B is the "electron acceptor," and the process is referred to as "electron transfer." The process is the same as tunneling. The tendency of two or more atoms or molecules to share, or thus "transfer," electrons is called their degree of "electronic coupling."

In general, the probability amplitudes for electrons "belonging" to widely separated atoms even in the same molecule don't merge in this way. For a long time, it was supposed that since electronic coupling falls off rapidly with distance, long distance electron tunneling could play little or no role in biological molecules—they're just too big. But recently this was found to be incorrect. Remember how, with the right set-up—e.g., a double-slit experiment—interfering electrons could show up anywhere at the detector, even very far away. It turns out that in proteins in particular, there exist very significant long-distance tunneling "currents."[6]

It first seemed that these "currents" must travel along the peptide string like an electric current along a wire. But this was an oversimplification. Certain protein structures called "bridges" (not peptide bridges) have the peculiar quality of enhancing electron tunneling along multiple, widely separated but nonetheless superposed paths, all with the same beginning and end. Figure 15–7 shows such a situation. It is more like the superposed photon paths in a double beam-splitting setup than in a double-slit experiment.[7]

In other words, for a time a portion of the protein is in a superposed state that involves substantial distances. Furthermore, superposed quantum paths can interfere. This is utterly impossible for pointlike electrons transferring along some kind of fixed "highway" and leads to effects not present in any classically mechanistic view of biological processes[8]—including unexpectedly long-distance tunneling. For certain bridge types, interference is both destruc-

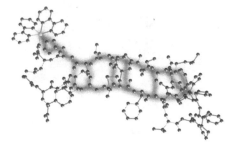

FIGURE 15-7   Long-distance electron tunneling in a protein. The blurred gray areas represent superposed paths. Note the "bridges" that greatly increase the number of possible paths from extreme left to extreme right.

tive and nonlocal—the presence of the bridge at one point causes destructive electron interference due to the presence of a distant bridge. (This is similar to how inserting a third detector in a beam-splitting gate destroys effects happening elsewhere.) When this happens, the electron tunneling process can become exquisitely sensitive to otherwise minor couplings: A certain far-away atom could donate but doesn't, because the probability is so low, yet an effect occurs anyway: Think of both quantum bomb detection and of computers that compute without computing.[9]

Different bridge structures and settings yield different influences on tunneling. For example, tunneling in helical secondary structures shows "a surprisingly weak decrease with distance," especially if it contains many residues of the amino acid proline.[10] This also means that the conformation of the protein as it folds influences whether, where, and to what extent tunneling occurs.[11] It works the other way around as well: The disappearance of an electron here and its reappearance there triggers conformational shifts—as in tubulin, whose "mobile" electron tunnels between a and b pockets in just this way.

TUNNELING AND CHIRALITY

There are situations where tunneling creates results that could not possibly happen otherwise. In fact, the most ubiquitous and important molecule to life—the one surrounded by which we came into existence and of which our bodies are by far mostly composed—that is, water, has been discovered to assume "chiral" (left- and right-handed) forms and to rapidly tunnel between them.[12] (See figure 15-8.)

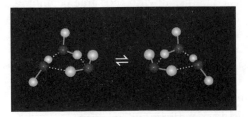

FIGURE 15-8   Tunneling of hydrogen creates mirror-image enantiomers of cyclic water clusters.

Without breaking the bonds and rearranging the molecule, *there is no set of rotations in a three-dimensional universe that could create a right-handed mirror image out of the left-handed version*. It's tantamount to "rotating" the molecule through a *fourth* spatial dimension. Water ordered in this way into chiral clusters plays an important role in protein folding.

Proteins evolved in an environment that allowed for and included all these quantum effects. At the least, terrestrial biological organisms must have adapted so as to ensure that these effects did not *prevent* the folding of proteins. We know this because they do fold, extremely well. Indeed, since these tunneling effects would open up shortcuts to many different, otherwise inaccessible states, it would increase the availability of low energy states among them, and these, being favored, would persist. In the end, with tunneling happening, low energy states would be attained routinely. So it's likely that life would have learned to use quantum effects, not merely to avoid their negative consequences. This, too, they evidently did, since quantum tunneling effects appear to be involved in the conformational changes required for enzyme-mediated catalysis.[13] Via the primordial genetic algorithm—evolution—life solved an extraordinarily complex problem in nonlinear dynamics, optimizing protein folding so as to both adapt to and take advantage of a quantum-influenced milieu. As we are now in a position to understand, the parallel computation that proteins perform is not quantum computation of the first kind, but of a second kind: quantum annealing of the sort we alluded to at the end of Part One.

BETTER VIBES

Quantum annealing does not require, nor would one expect to find associated with it, the perfect isolation from the environment without which quantum computation of the first kind is utterly impossible. Nonetheless, it is generally the case that any process which benefits from pseudorandom agitation (heat) benefits little from tunneling: The heat-induced bouncing about interrupts tunneling and overwhelms whatever of it is left. In proteins, however, just the opposite turns out to be true, making even stronger the evidence that living matter takes direct advantage of quantum effects to do what life alone does.

Just as there is solid evidence for *transient* long-distance superposition of regions of a protein's electron cloud, there is likewise evidence for *transient* propagating *regions* of classically coherent sound waves in proteins—*phonons*: "particles" of sound, if you will. (A tsunami is a gargantuan phonon.) When the equations for such localized but traveling forms of classical, mechanical coherence in proteins are calculated and added to the equations that govern the quantum "motion" of tunneling electrons, it turns out that, *because of the feedback between tunneling and conformational change, phonons enhance tunneling distance and frequency*[14]: ". . . the quantum mechanical tunneling process . . . for electron transfer involves in an intricate way a nonlinear dynamic of the protein medium in which the tunneling occurs. It is the couplings of electron and vibrational degrees of freedom (phonons) that seem to be of critical importance for the dynamics of electron transfer in proteins."[15]

Tunneling in turn affects the extent of phonon formation (in an especially dramatic way, to be clarified shortly).[16] In short, proteins have figured out a way to create a mutually reinforcing relationship between classical mechanical jiggling and quantum weirdness so as to amplify their powers far beyond what would be possible were they the mere molecular machines they are still widely thought to be. Physiological jiggling *precludes* single, stationary, long-lived, large-scale quantum coherence (à la Penrose), but it *enhances* multiple, migrating, short-lived, small-scale coherences, which, in honor of Max Born, we may call . . .

QUANTUM GREASED LIGHTNING*

By superposing itself in (technically) an infinite number of paths—*all* possible ones—*an electron is able to search simultaneously for the most efficient one.* The protein responds to the electron shifts by changing its conformation. The new conformation, in turn, alters the probabilities that determine the different paths and alters as well their relative efficiencies. While a completed conformational change of a whole protein may take seconds or minutes, even tiny changes—on the scale of a vibration—can occur as good as immediately, and these, we have noted, have critical effects on the tunneling process. There is, in other words, a continuous feedback process that is as close to instantaneous as one might imagine. To a degree, then, we may speak of "parts" of the protein folding process that are greased to lightning speed by the quantum fuzziness.[17]

Recently a research group headed by Peter Wolynes at the Center for Biophysics and Computational Biology at the University of Illinois Urbana-Champaign was able for the first time to simulate the tunneling process in protein: "We have successfully shown that the theory of spin glasses and of neural networks can be applied to . . . solving the protein folding problem.[18] Using Feynman's path integral formulation of quantum mechanics, the electron tunneling act can be thought of as arising out of a superposition of many possible paths. . . ."[19]

At the National Center for Supercomputing Applications, Wolynes was able successfully to model the 800,000 most important superposed paths for one protein. (Figure 15–7 shows a handful of hypothetical paths.)

At the end of Part One, in our discussion of quantum annealing, we noted that "if tunneling between states . . . is allowed in a system, its computational function is strikingly enhanced." This was tried in reality on a "Coulomb Spin Glass" minimization—which happens to be especially closely related to protein folding[20]—with results that were a hundred times more efficient than without tunneling. In a related problem, tunneling was also far more likely to allow a system to anneal to its *absolute* energy minimum than was annealing without tunneling. Proteins achieve lower energy states not

---

* Why Max Born? It was he who first recognized that quantum mechanics implies the existence of absolute chance and required the abandonment of Newtonian causality. It is this that allows for the fantastic speedup and efficiency of living processes. His granddaughter is Olivia Newton-John, of *Grease* fame.

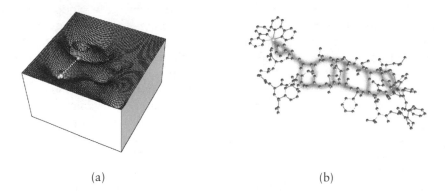

(a)   (b)

FIGURE 15–9   Direct access of low energy state via quantum mechanical tunneling. (a) State represented as the location of a ball in an energy landscape ("energy space"). (b) Tunneling paths represented in physical space, as in figure 15–7.

merely by an abstract tunneling of "states" in an energy landscape, but by the physical "teleportation" of electrons, if you will, from one point in space to others. (See figure 15–9.)

In proteins, therefore, we see that nature has learned to take advantage of environmental intrusions, even perhaps of decoherence itself: Each lightning-like burst of superposition decoheres into just the right nonsuperposed state so as to set up the protein for the next burst. A sequence of events, seemingly orchestrated, ensues to leave the protein properly folded so quickly as to be utterly beyond imagining in a deterministic universe. This process occurs ceaselessly in every living being, in trillions upon trillions of proteins within every one of the trillions of cells that comprise it.

ELECTRONS DO IT, PROTONS DO IT, TOO

Hydrogen atoms in proteins, with their nucleus of a single proton three orders of magnitude larger than electrons, also tunnel long distances, and this, too, has been found essential to protein function, in particular to enzymatic action. Again, as the protein changes, its structure alters the proton tunneling, and the proton tunneling influences the conformation of the protein.[21]

Two recent discoveries involving hydrogen atom tunneling are especially dramatic. A group at the University of Leicester found that not only are "extreme" proton superpositions possible in proteins in spite of environmental vibration, environmentally induced vibrations are *required* for it to happen. "Extreme" tunneling refers to tunneling from the lowest energy ("ground") state—normally the most difficult from which to tunnel since objects in this state are the least delocalized. External heat sets up internal vibrations in the protein, and protons in the ground state use the vibrations to tunnel to distant locations.[22] In fact, "at long distances, the phonon[vibration]-modified . . . tunneling always dominates over . . . [spontaneous] tunneling."[23]

In other words, since tunneling is itself an intrinsically random process, merely by absorbing heat—which is to say, disorder—from the environment, proteins are able to initiate orderly computational processes spontaneously, without being "activated" to do so by some directed energy. Like pulling oneself up by one's bootstraps, life appears to be able to use quantum effects as a kind of "ratchet," extracting order and directed activity out of disorder. This is at least one of the sources of its extraordinary efficiency, which applies not only to protein folding but to more general living processes, as well: "Molecular motors are single protein molecules that put random Brownian motion to use. For instance, molecular motors drive the contraction of muscles and carry out intracellular material transport [one of the functions of the microtubule network]. The extremely small size of protein molecules make it likely that they use quantum effects to perform their tasks, but understanding of protein motors is only just at the beginning."[24]

Very recent research tends to confirm directly that life has figured out extraordinarily ingenious ways to incorporate quantum effects. In June of 1999, Judith Klinman's research group at Berkeley published a study on an alcohol-metabolizing enzyme found in a heat-loving organism, *Bacillus stearothermophilus,* that thrives in temperatures around 65 degrees C (150 degrees F). In general, the proportion of proton transfer in enzymes that occurs via tunneling goes down as the temperature goes up, being most prominent, therefore, at low temperatures—as with most quantum phenomena ruined by random agitation. Earlier research by her group showed that significant hydrogen tunneling nonetheless occurs in proteins at room temperature, as with electrons.

But in the *Bacillus stearothermophilus* alcohol enzyme, hydrogen tunneling actually *increases* above room temperature. The creature's tunneling capacity has evidently evolved to take advantage of the unusual ecological niche to which it has adapted[25]—close by oceanic volcanic vents:

> Our present findings on hydrogen transfer under physiological conditions cannot be explained without involving both quantum mechanics and enzyme dynamics.
>   . . . The implication of [vibrationally enhanced tunneling] protein dynamics in modulating chemical steps makes it easier to understand why . . . [standard] models . . . often fall short of the rate accelerations observed.[26]

Proteins even seem to have learned how to take advantage of proton tunneling in the immediately surrounding molecules of water itself—tunneling, we have recently learned, is apparently ubiquitous—including even the tunneling transformation of left- handed into right-handed clusters, just as if being turned through a fourth dimension.[27]

EVOLUTION AS A QUANTUM GENETIC ALGORITHM?

Clearly, these phenomena are not likely to have arisen just by happenstance. They are the result of an adaptive process that evolved tunneling-sensitive

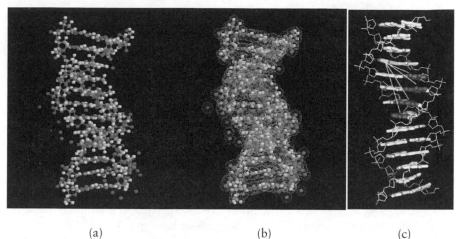

(a)   (b)   (c)

FIGURE 15-10   Segment of DNA. (a) Ball-and-stick representation of nuclei and bonds. (b) Mean radii of electrons. (c) Electron transfer in DNA. Straight lines show origin and destinations, not pathways.

structures—for example, cross-bridged helices.[28] But proteins are not the only biomolecules with bridged helices. DNA is surely the most famous of these. (See figure 15–10a and b).

In a series of exquisite experiments over the last few years at Stanford, Jacqueline Barton and colleagues demonstrated that electron tunneling in DNA occurs across as many as sixty base-pairs:

> The implications of long-range charge migration through DNA to effect damage are substantial. . . . [T]hese reactions depend . . . upon the integrity of the intervening base pair stack, *but not upon molecular distance.* [See figure 15–10c.]
>
> Also, a physiologically important DNA lesion . . . can be reversed in a reaction initiated by electron transfer, and this repair reaction too can be promoted *from a distance.*[29]

This means that both normal evolution (involving mutation and crossover) and cancer (involving DNA lesions) are in part nondeterministic processes, influenced by "causes" that can be found nowhere in the physical universe. Furthermore, the repair process referred to as being "promoted from a distance" *would be not merely "uncaused" but absolutely impossible without quantum effects.* In an interview, Barton concluded: "We are talking about biologically relevant distances, and you can have strange fantasies about what the implications might be."[30]

In other words, the claim that "anything can be reduced to simple, obvious, mechanical interactions. The cell is a machine. The animal is a machine. Man is a machine"[31] is looking increasingly dubious. At least, the stuff of life is a "machine" only in the way that quantum mechanics is "mechanical"—shot throughout with uncountable numbers of uncaused events, possibly taking place in a nearly infinite number of cross-talking parallel universes.

At least one group has in fact been considering the possibility of proton tunneling in DNA as well and has raised the question of a strange mutual interaction between various illnesses and evolution—making of them both a single, quantum-mediated process: "The model further illustrates how [certain genetic illnesses] could be a result of evolutionary lesions altering genetic specificities of '[proton] tunneling sensitive' . . . codes."[32]

Some other "strange fantasies" that may soon no longer appear so strange include the speculation that illnesses which are "information rich"— HIV (human immunodeficiency virus), for example, which interfaces with the immune system (another major model of the computational paradigm in biology)—are especially dense in quantum events.[33] From this point of view, the battle between the body and AIDS—because of how quickly HIV mutates and adapts to whatever the immune system throws up against it—is more like a battle of miniature magicians than of microscopic machines: a war of adaptation waged by opponents with equal access to quantum opportunity.

A few serious researchers have gone considerably further in their speculation. For example, Michael Conrad, head of the Biocomputing Group at Wayne State University, argues that "proton superflow"—a dynamically fluctuating superposition of all hydrogen atoms—is ubiquitous in proteins.[34] It's far less speculative than the 20,000 cell superposed microtubule network of Penrose and Hameroff but far more than what most scientists would today entertain. Yet at the time Conrad proposed it (1990), the idea of even single proton superposition and tunneling in biomolecules raised eyebrows.

BACK TO BOB

Bob, we learn, has married Alison. With her "yes," they have thereby set in motion a unique chain of events that was never predetermined. We have thus proven that the human brain amplifies quantum effects into the scale of terrestrial life—at least certain human brains do, Bob's preeminent among them. Given the scale of the universe, does it really matter whether the scale at which the amplification occurs is that of atoms or that of grapefruits?

The philosopher's answer is "No, it doesn't matter." Bob's grapefruit-size brain is a "black box" to us. That Bob's brain is a quantum amplifier is evident. Is it a mechanical machine? Then mechanical machines lacking any freedom whatsoever can incorporate quantum effects and thus generate otherwise undetermined outcomes, and so are no longer mechanical. Is Bob's brain not a mechanical machine to begin with? Either way, the matter is settled: It's not now.

But there's something a bit too slippery about this "black box" argument. It would be awfully nice if we could show that the quantum computation of the second sort going on all the time in proteins—in tubulin, in particular—is directly amplified upward by known means. To show this we have to answer two questions. The first is: "Are there any means (other than a person's deliberate manipulations) by which tiny quantum effects can be amplified upward in scale to everyday life, instead of just being averaged away?" If the answer is "yes," and we can say what they are, we must then answer a second question: "Is there any evidence that such means are present in and employed by the human brain?"

There are good reasons to suspect that the answer to both questions is "yes." As to the first question, in general, systems that are prone to chaos routinely take differences of very tiny scale and amplify them upward in scale. In order for a system to be capable of chaos, it must be iterative, that is, characterized in the main by internal loops and feedback such that small, local effects spontaneously generate large-scale patterns. In order for some very tiny difference at some very small scale to be amplified upward across many scales, with possible different kinds of structures at many of the intervening scales, the demands will be even more specific: The main structure at every level must be iterative in one fashion or another. In this way, the initial very tiny differences generate a unique pattern at some intermediate scale, which serves as a small difference within a single element among many like it at a larger scale, which is itself amplified upward, and so on.

As to the second question, we now know that the above is precisely how the human brain is structured. It needed to be structured this way, at every scale, so that it could operate as a self-organizing parallel computer, one that generates global order out of purely local effects. But precisely those features cause it also to display chaotic potential at every scale and thus amplify upward—not average away—differences that arise as a consequence of yet smaller differences at the next scale down. The great "uncaused cause" that theologians have long speculated lies both everywhere around us and also within us, and of whose freedom our own is claimed to reflect, is not so bad a description, after all, of what may actually be the case. Let's see how it works.

16

TO FICKLE CHANCE, AND CHAOS JUDGE THE STRIFE

*Either it is a well-arranged universe or a chaos huddled together,
but still a universe. But can a certain order subsist in thee, and
disorder in the All?*
—*The Meditations of Marcus Aurelius, IV:27*

At every scale, from the cortex as a whole down to individual proteins, the matter of the human fourth brain functions as a parallel processor of one kind or another. These processes form a nested hierarchy, an entire parallel computer at one scale being but a processing element in the next larger one. (See figure 16–1.)

We have focused on the unique features of the processing at each scale as well as on how they are variations of a common theme: self-organized parallel computation. All forms of parallel computation require *iteration*: Data in parallel processors circulates round and round. Feedback loops of varying sophistication and complexity form the structures of every brain, from first to fourth, and form *oscillators*. Input goes into an element, it's changed according to some rule, is perhaps fed out to other units where it's changed some more, and eventually returns in altered form to the original unit, where the cycle begins again.

The effects of iteration at one scale on the behavior of larger-scale aggregates are not what science long thought. Of course, *all* matter is organized by scale: A static lump of rock has large-scale volumes bounded by fracture planes; these volumes are composed of disordered regions of crystalline mineral; each region is composed of many aligned crystals; the crystals, of molecules; and so on. A similar hierarchy of scale exists for dynamic systems—water flowing in a stream, for example. In such dynamic systems, the motions of elements on the smallest scale combine to create patterns on the largest. Even were it possible to derive the large-scale dynamics by literally adding up the individual motions of every one of the smallest elements, it would be a waste of time. By treating the small-scale actions *statistically*—by assuming, in other words, that most of their actions average out—science has arrived at ex-

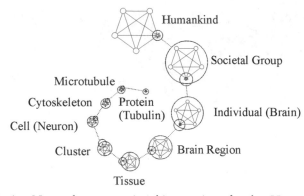

FIGURE 16-1   Network computational integration of scales. Humankind occupies a volume that is roughly 8×10^6 cubic meters; a protein occupies a volume that is roughly 6×10^{-26} cubic meters. The scale difference is ca. 32 orders of magnitude (10^{32}). Star-shaped nets represent any form of distributed parallel processing.

tremely precise approximations for the dynamics of many systems composed of large numbers of smaller elements.

CLASSICAL CHAOS

But only in the last century, and in the last quarter century especially, did science learn that for iterative systems, the statistical averaging falls apart. In iterative hierarchies, not only are small differences not washed away, they are usually amplified. It was a shocking discovery, and it gave birth to an exploding new field of mathematics and science—chaos theory. Nor is chaos theory required only when studying exceptional situations. Quite the contrary: "[N]onchaotic systems are very nearly as scarce as hen's teeth, despite the fact that our physical understanding of nature is largely based upon their study."[1]

Consider the self-organizing spin system we discussed before (reproduced in figure 16–2). It began in two very different initial states and evolved iteratively as the spins interacted. The two final states were also very different: There was only a 50:50 chance that any given cell (spin) would have the same color (spin direction) in both runs. We emphasized the fact that even so, the overall patterns were recognizably similar. The outcomes were thus "ordered" but within that order highly variable—like different individual cows but of the same Guernsey breed.

It might seem that the large-scale differences in the final states result from the *many* individual small-scale differences in the initial state. But this is not necessarily so. Because the evolution is iterative, *even very tiny initial differences generate equally large differences in the final state.* In figure 16–3 the same spin simulation program is used to iterate two initial states that differ by just one spin in 1600 (the horizontal arrow identifies the line in which one spin has been flipped)—an initial difference of one-twentieth of 1 percent—yet again, we see that in the end, there is only a 50:50 chance that any given cell will have the same color in both runs.

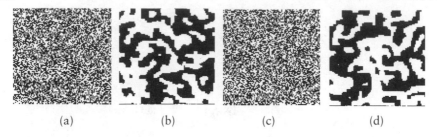

(a)   (b)   (c)   (d)

FIGURE 16–2   Spin evolution from two very different initial states, (a) and (c), to two very different, yet similarly patterned final states, (b) and (d).

FIGURE 16–3   Spin evolution from two very similar initial states (left) to two very different, yet similarly patterned final states (right).

Such "extreme sensitivity to initial conditions," as it's called, is a hallmark of iterative systems, and nowhere in the known universe exists a structure more densely iterative than the human fourth brain. Dense iteration gives the brain its enormous computational capacity, as we've seen, but it also means that the way the brain handles the transition from the extremely small to the everyday is going to be entirely different from a noniterative, unintelligent system the dynamics of which conform to the usual statistical approximations.

In the brain, self-organized parallel processing (hence, iteration) takes place at each scale, is influenced by the adaptational requirements of the scale above it, and in turn exerts influences on the processing taking place at the scale beneath it. The final outcomes at each scale constitute the initial states of the individual elements at the scale immediately above. Since "initial state" is an arbitrary designation—*any* state is "initial" to the ones that follow—these relations between scales unfold continuously.

The fourth brain could hardly have been better designed to amplify quantum effects upward. This amplification, however, does not allow one to witness large-scale aggregates of living matter obeying quantum rules, as pop quantum mysticisms propose.[2] To all appearances, the mechanical model holds relentless sway. And yet it is possible to see that, in the end, it doesn't.

PLAY IT AGAIN . . . AND AGAIN, AND AGAIN . . .

Figure 16–4 shows the world's simplest possible iterator, shown on the left as the simplest imaginable neural network—a single neuron—and the simplest imaginable cellular automaton—a single cell.

FIGURE 16-4   A simple iterator (left) as a neural network and (right) as a cellular automaton.

The neural network version has a "transfer function" that takes the input—which happens to be its own output—operates on it, and puts it out. If we think of its output as its state, then depending on what the initial state is and the specific transfer function, at each iteration its state will (or may) change. The cellular automaton version has a "transition" or "state change" or "update" rule. Since its neighborhood is simply itself, it takes its present state, runs it through the rule, and changes accordingly. At each iteration its state will (or may) change. If the transfer function on the left is the same as the update rule on the right, and if the initial states are the same, too, then the successive states of both will obviously be identical. These are merely two different visualizations (based on biology) of the same mathematical process of iteration. So, let's invent a transfer function/update rule (which we'll call the "rule"), choose some initial values, and see what happens. We'll label the input upon which the function is going to operate at any given iteration x. We'll label the resulting output x'. This means that on the next iteration, a new x is going to be used that's simply the previous x'.

Now, suppose the rule is an ordinary parabola, $x' = ax - ax^2$ (where we're used to seeing y in place of x'). On the left in figure 16-5 is a graph of this parabola as an ordinary equation, with $a = 4$, for values of x between 0 and 1. (In case you're used to seeing U-shaped parabolas like $y = x^2$, the a narrows it, the minus before the ax^2 turns it upside down, the ax changes its shape and moves it up and over.) On the right is an *iterated* graph, where we "seeded" the rule with an initial value of $x = 0.5$

With $a = 4$ and $x = 0.5$, the iterated version first puts out $x' = 1$; when 1 is fed in, it puts out $x' = 0$; and when 0 is put in, it again puts out 0. Hence, it attains only two values, both of which are found on the underlying parabola (shown in gray). But after the first value it heads to 0 and stays there—that's why it's circled. For this rule, and these values of a and x, whether imple-

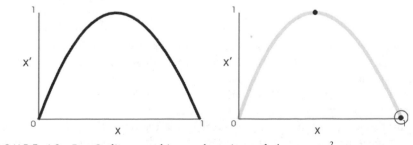

FIGURE 16-5   Ordinary and iterated versions of $x' = ax - ax^2$.

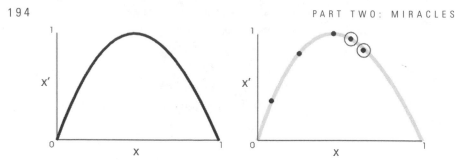

FIGURE 16–6   Iterator with two stable attractors (circled).

mented as a cellular automaton, or as a neural network, or just mathematically, 0 is an *attractor* (its only one), as in *basin of attraction.* (Even this simple
example suggests the relationship between iteration and the formation of energy basins.)

Now let's set $a = 2$ with the initial value of $x = 0.05$. This time, successive
values of x' are 0.05, 0.09, 0.17, 0.28, 0.41, 0.48, 0.49, 0.5, 0.5, 0.5. . . . After
a few iterations, it's again attracted to a (different) stable value, namely 0.5.

A more interesting case is shown in figure 16–6 where we let $a = 3$ and
start with $x = 0.1$. This time, after a few iterations, successive values of x' settle
into an alternation between *two* stable attractors, 0.71 and 0.62 (circled).

In other words, with this value of a, a cellular automaton or neuron with
the above rule has *two* final states: It has become a stable, two-state *oscillator*
(like an ammonia molecule with two equally likely low-energy states). Furthermore, it turns out that with this particular rule, if we always keep our starting
values between 0 and 1, then as we increase a from 0 to about 3.57, the rule
starts out producing first one final state, then suddenly two final states, then
four, eight, sixteen, and so on. This abrupt doubling of the number of final
states happens at ever *smaller* increases in the value of a. The *ratio* of one increase in a between doublings to the next increase is always the same, a number whose inverse is known as *Feigenbaum's constant.* No one knows why this
is so. Furthermore, this ratio is small enough so that before a reaches 3.58, the
doublings reach infinity.

This means that above a certain value of a, the value of the "oscillator"
no longer oscillates in the usual sense—its values never cycle through the exact
same place twice, even though they are forever *bounded* by 0 and 1. It looks
random but isn't; its values are 100 percent mechanically determined, forever—by the rule, by the value of a, and by its initial value. This phenomenon
is known as *deterministic chaos,* as we mentioned briefly at the beginning of
Part Two.

Now let's graph the value of the final states of the oscillator against a few
values of a (see figure 16–7 left) and for all the values of a between 0 to 4 (see
figure 16–7 right).

We see the entire range of final-state doubling and the onset of chaos—
each point in the speckled gray regions represents a value of the iterator after a
large number of repetitions. If we now enlarge a region of this "final-state diagram," then take a similarly sized region from the enlargement and enlarge

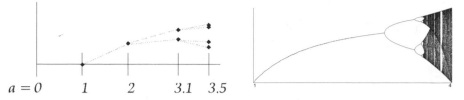

FIGURE 16-7   (left) Final state diagram for a quadratic iterator at discrete values of *a* between 0 and 4. (right) Final state diagram for a quadratic iterator for values of *a* between 0 and 4.

FIGURE 16-8   Scaling of the transition region from a doubling regime to a chaotic regime in an iterator at different "zoom-in" values. In the leftmost diagram, *a* ranges from 1 to 4, in the next from 3.0 to 3.7, in the next from 3.45 to 3.59, and in the rightmost from 3.544 to 3.575. The scale of the rightmost range is therefore about 1/100 of the leftmost.

that, and keep doing this, we find that at every scale, the pattern repeats itself in finer and final detail (but inverted from the prior scale—see figure 16–8).

I mentioned that a chaotic sequence of values looks random but isn't, because it's mechanically determined. It's also the case that within boundaries, and in spite of never repeating the same value twice, obviously related overall patterns appear that are distinct for the given rule. Furthermore, *any two very slightly different initial values will soon generate totally different sequences of values.* This is just what we saw with the spin system in figure 16–3. The specifics differ dramatically as a result of a tiny initial difference, but the general pattern remains recognizably the same.

Now, figure 16–9 is a graph of the first 1,000 values of our iterator.

At a glance, there seems no evident order to the values. And yet we see that the values are sharply bounded and within those bounds there are also bands of approximate order—aggregates, for instance, near the top of the range and the bottom, and a kind of band in the middle. If we look at widely separated sequences of iterations, we see forms that have about the same shape but do not repeat themselves exactly, as in figure 16–10.

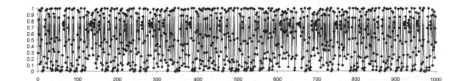

FIGURE 16-9   First 1,000 values of an iterator.

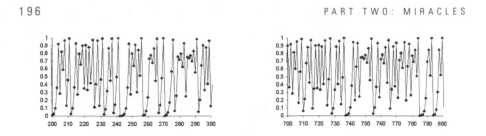

FIGURE 16-10   Widely separated regions of an iterator. (left) iterations 200 to 300. (right) iterations 700 to 800.

In a (very) loose sense, we may speak of the individual values as being "random" but distributed according to larger rules that govern the probability of finding a point in any given region. This may sound a lot like quantum formulations, but *be careful*. A quantum event *is absolutely random*, determined by nothing in the universe. Chaos, however, *is absolutely determined*. If you know the rule and the starting value, you can predict the results to whatever degree of exactness you wish for as long as you wish. In a quantum regime, even in the theoretical extreme where you know everything there is to know about it, you absolutely cannot predict the outcome—just the probabilities of various outcomes.

THE FLAP OF A WING

Chaotic (iterating) systems are *extremely sensitive to initial conditions,* and because they never repeat, they are impossible to predict unless you know all the underlying equations perfectly and the present (initial) condition with perfect precision. You can be off by one part in a billion, and very quickly your predictions diverge to the point of meaninglessness. The first object for systematic study of chaos was the weather, notoriously unpredictable because, as the title of the classic paper by Edward Lorenz asks, "Can the flap of a butterfly's wing stir up a tornado in Texas?"—the answer being "yes." Therefore, while general patterns remain the same at every scale, an iterative system will amplify upward small individual differences at the lowest scale so as to generate massive global differences at every larger scale.

COUPLED OSCILLATORS:
STRANGELY ATTRACTED TO YOU

Let's keep these ideas in mind as we proceed: So far, the neural network/cellular automaton system we've examined is simple to the point of silliness—one neuron, one cell. Even so, it demonstrates all the essential features of chaos, the true mathematical complexity of which we've avoided. Imagine how much more complicated things become when even *two* such oscillators influence each other; or ten; or 1,600 as in figure 16–3; or 20 trillion, as in the neural network of the brain. Especially if the individual elements are not perfectly

FIGURE 16–11   Iterated graph of $u' = w + 1 - au^2$.

identical, but have complicated, slightly different rules (as will happen in any natural system in a noisy environment), it *seems* intuitively obvious that when coupled, the fuzzy patterning of a single chaotic system will be washed out beyond recognition, but this is not necessarily so.

For example, we can couple two neurons (or two cellular automata) to each other by coupling their rules, which now contain two variables, say u and w, and two parameters, a and b. Since the math becomes vastly more complex immediately, we're not going to go into as many specifics as with a single iterator, but here's an example:

$$u' = w + 1 - au^2$$
$$w' = bu$$

Note that in this (deliberately simple) case, the new value for u depends on the previous values of both itself and of w, and the new value of w depends on (only) the previous value of u. In spite of the simple appearance of these equations individually, the coupling between them makes their behavior extremely complex. So let's ignore w and just make an iterative graph of u. That's like tracking the state of just one neuron in a network of two neurons. Let's jump immediately to a set of values for a and b such that the coupled system is chaotic—it never attains the exact same value twice and therefore never settles into a stable attractor. Figure 16–11 shows about a hundred states for u', graphed against u.

That this conforms to a pattern is obvious—more obvious than the jumping about of our single iterator. If we continued to plot more points, we would discover that, however much they jump, they form seeming lines that continually *fold* back and forth but never cross (like the folded sheets of very fine French pastry).

A conventional attractor is like a low-energy basin with one (or more) exact lowest position(s). When a system is chaotic, it will no longer reach any such fixed position(s). Indeed, as we've noted, as it evolves it never touches the same place twice, let alone remain there permanently. However, we see that chaotic systems tend to hover in the neighborhood of one or more general regions, much like frustrated spin glasses. These "fuzzy" basins are called *chaotic strange attractors*. The pattern in figure 16–11 is a famous example of such an attractor.

FIGURE 16–12   Four strange attractors generated by two coupled oscillators with two variable parameters.

Many processes in nature that look thoroughly random in fact have an underlying "very high dimensional" chaotic dynamic due to the reciprocal coupling of large numbers of individual elements—each of which itself might have a completely nonchaotic dynamic (such as would result from iteration of a straight-line equation), or a simple chaotic one (such as results from iteration of a parabolic equation), or a complex chaotic one, as would result from the iteration of many reciprocally coupled chaotic systems each of which is composed of many reciprocally coupled chaotic systems, and so on.

Even relatively simple systems of coupled oscillators can generate an astounding variety of chaotic, but obviously ordered, patterns. Figure 16–12 shows four that all come from two coupled oscillators with slightly varying parameters a and b.

Even though, as in the one-dimensional case for x alone, neither u nor w ever take on values that repeat, the values cluster in obvious regions in all four cases.

The fact that globally defined *patterns* are repeated with variable, but never perfect, degrees of precision while individual values are never repeated is perhaps in no way more dramatically illustrated than in this example offered by James Crutchfield in his famous 1986 *Scientific American* article that put chaos theory on the map.

A PORTRAIT OF POINCARÉ

A picture—in this case of the great French mathematician and physicist Henri Poincaré—is subjected to a transformation that consists of a series of (mathematically expressible) operations. These distort both its boundary and its interior. When the transformation is finished—the equivalent of subjecting an input variable to a series of mathematical operations—the end result is a new picture of the same dimensions as the original. The process is then repeated, with the output of one transformation becoming the input for the next iteration, as illustrated in figure 16–13.

We may consider the original portrait to constitute a "position" for a large number of variables—the spatial location of each pixel, for example. If the principles of deterministic chaos hold, we expect at least some future iteration to come back again to a "point" in this multidimensional "space" close to but not exactly at the starting point—that is, something close to the original arrangement of pixels. Rather amazingly, that is just what happens—repeatedly—a phenomenon known as Poincaré recurrence. A sample is shown in figure 16–14.

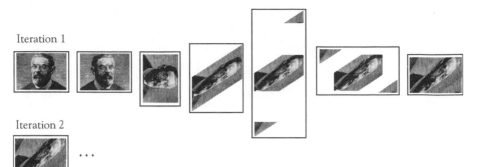

Iteration 1

Iteration 2

...

FIGURE 16-13   Transformation of a portrait of Henri Poincaré. One complete iteration has five steps (shown in seven pictures) from left to right: (1) horizontal mirror, (2) rotate 90 degrees, (3) vertical skew, (4) make a lower right corner of the upper
right corner, (5) make a lower left corner of the upper left corner.

In nature, many iterative processes are of such high "dimensionality" (iterative nests within nests within nests . . .) that recognizably close recurrences
are as rare as a blue moon. Furthermore, as the number of iterations increases,
Poincaré recurrences of a given level of accuracy on average grow ever more
rare.

Iteration 1   Iteration 2

Iteration 5   Iteration 18   Iteration 48

Iteration 237   Iteration 239   Iteration 241

FIGURE 16-14   Iterated transformations of a portrait of Henri Poincaré. By iteration 18, one would assume that nothing of the original information about the location
of pixels was retrievable. Yet at iteration 48 a multiplied version appears and by 241
an almost (but not quite) perfect recurrence. These recurring patterns are known as
Poincaré recurrences after the man who first discovered them.

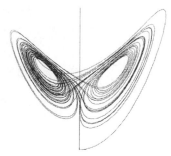

FIGURE 16-15   Lorenz attractor, a three-dimensional iterative strange attractor.

Two important facts to store for use shortly are that *in a deterministic chaotic system, the fall-off in the frequency of Poincaré recurrences is "exponential,"* that is, very rapid*; but surprisingly, *in a deterministic chaotic system formed out of a nested hierarchy of lower-level chaotic elements, the fall-off in the frequency of Poincaré recurrences is only algebraic,* that is, slower.** Recurrences are *more* frequent if the overall system is composed of a nested hierarchy of chaotic subsystems—as is the brain.

From such examples as the well-known three-dimensional "Lorenz attractor" (first discovered in an investigation of weather patterns, see figure 16–15), an important point can be made about strange attractors in general (energy basins for chaotic systems) and the phenomenon of Poincaré recurrence.

Notice that the attractor is composed of two orbits, or "wings." A system that conforms to an attractor like this is apt to be found in one of two distinct, dichotomous basins of attraction. *Which* of the two basins it will be found in at any point in time, however, is dependent on its initial state. Thus, two systems identical to a thousand decimal places—but differing there—will as commonly end up after, say, a million iterations (probably far fewer, in fact) somewhere in the same basin as in the other: tiny initial difference, large-scale dichotomous result.

In other words, suppose you have two absolutely identical twins—far more identical than biology actually allows: that is, identical down to but a single amino acid in one protein in one neuron, and raised (purely hypothetically) in *absolutely* identical environments. Is it possible that in certain domains, they will turn out *opposite* rather than identical? The answer is "yes," for the above reasons. Then how do we explain the great similarity of even real identical twins? By the fact that the at-a-minimum dichotomous nature of iterative outcomes is nonetheless strangely bounded, as well, yielding outcomes that however divergent in the individual event conform to broad patterns outside of which they *never* fall.

---

* $\frac{1}{p^t}$, where t is the number of iterations and p is some positive number greater than 1.

** $\frac{1}{t^p}$, where p is typically ~15.

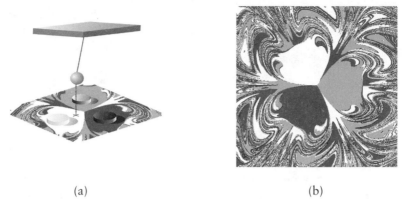

(a)   (b)

FIGURE 16-16   (a) Pendulum in three magnetic fields. The location of the pendulum in the plane of the magnets is marked by X. (b) Final positions of the pendulum coded in dark gray, gray, and white.

CHAOS AND SPIN

A simple physical model closely related to the phenomenon of coupled oscillatory neurons (or competing spins) is an iron ball pendulum set swinging over three disc-shaped magnets (white, gray, and black), as in figure 16–16a.

There are three basins of attraction for the pendulum, one nearest each magnet. But the relationship between the initial location of the ball and its final spot is anything but simple. The forces acting on the ball depend on its height as well as its position over the magnets, and these all change with its speed and direction of travel (that is, its momentum). Furthermore, it is an intrinsically oscillatory system. The actual relationship between initial and final states can be shown as the map in figure 16–16b. It is to be read as follows. Pick a point on the map. This is the ball's initial position above the plane of the magnets (which will automatically fix its height there). The color at that point on the map corresponds to the one magnet it will come to rest closest to.

Magnifications of this diagram reveal a fractal structure to the pattern at endlessly fine scales. In other words, you can always find some region of the map that contains all three final positions, no matter how small a region you choose. Thus, the system exhibits extreme sensitivity to initial conditions. In fact, the "small" initial differences can be infinitesimally small.* But there's an important subtlety. There are only three distinct final states. The system does not *evolve* chaotically—it comes to rest in a fixed basin of attraction, like a Hopfield net remembering a distinct memory. Chaos—small initial difference, large trichotomous outcome—is present in the structure of the boundaries between initial states. Thus, the final outcomes of the system—what we see— look as if the dynamic processes that produced them must be rather simple. In

---

*   Intuitively, it's fitting that such systems, where infinitely small differences can produce starkly opposite results, should require rules that have imaginary numbers in them, as such "Julia sets" do. Yet they describe real physical systems.

reality, they are chaotic and are exquisitely sensitive to initial differences, no matter how small. Indeed, small initial differences in iterative systems—even infinitely small ones—can have exactly as profound an effect on outcomes as do large initial differences, no matter how large!

LIFE, THE BRAIN, THE CORTEX

There is direct evidence that the structure of the brain is iterative and its function prone to chaos: The electroencephalogram (EEG), a relatively crude measure of brain electrical activity, turns out to demonstrate a remarkably consistent pattern of structured chaos, once the proper mathematical tools are applied.[3] Furthermore, when the brain is modeled by coupled oscillator neural networks with elements that incorporate as much as possible of what is known about real neurons,[4] the "simulations . . . show good correspondence with [the] . . . EEG."[5] Both the real EEG and the trace generated by a neural network model turn out to be fractal in nature with similar dimensionality.

Walter Freeman at the University of California and his group have developed a complete model of the human olfactory cortex, illustrating how both it and the actual cortex recall smells through an associative, attractor-type (Hopfield) neural network, with each smell corresponding to a basin of attraction. With each inspiration-exhalation cycle, the system alternates between chaotic and nonchaotic processing modes.[6] Other researchers have now succeeded in implementing a far more complex model that involves 49,000 artificial neurons with more than 100 million connections and the kind of distance-dependent signaling delays that are found in actual neural dendrites and axons. It shows dynamics even more strikingly similar to that of the actual human olfactory cortex.[7]

In the real world—in the brain model we have been developing—at any level the interior structure of each coupled element in a network is itself at the next smaller scale. Such an element is still called an "oscillator," but it oscillates among regions near a strange attractor. It seems eminently plausible that if you couple a small set of such individually chaotic "oscillators," the operation of the system as a whole should end up as close to random as you can get—totally unanalyzable, in any event, especially if you're speaking of thousands or *billions* of elements. But from the fact that Poincaré recurrences are *more* frequent when the oscillators themselves are chaotic than when they're not, we suspect that our intuition is going to fail us once again. It does, indeed. A network of even *chaotic* oscillators can enter into a state in which there is both global synchronization and network-level chaos especially useful "[f]or a description of . . . a number of biological phenomena."[8]

CHAOS IS A FRIEND OF MINE

Neural networks that are modeled closely on biological systems inevitably demonstrate chaotic dynamics.[9] For many years the presence of chaotic dynamics in an artificial neural network was considered undesirable, mostly because it made analysis of its behavior so difficult—a sufficiently complex chaotic system is very difficult to differentiate from a random one, even though it is now understood that such systems do have an underlying order. (The point was best made by John von Neumann in 1951: *"Anyone who considers arithmetical methods of producing random digits is, of course, in a state of sin."*)[10] Since chaos is ubiquitous in natural systems, and especially in such densely iterative systems as the nervous system (as in their spin glass, neural network, and cellular automata simulacra), it was proposed that there must be something beneficial about it.[11]

In September 1998 a group at the Applied Chaos Laboratory of the Department of Physics at the Georgia Institute of Technology showed that an array of coupled chaotic elements can perform *any* logical operation. That this must be so they had concluded simply by reasoning that were it not true, natural selection should have eliminated it: "[W]hile it is known that coupled chaotic maps can, *in principle*, be viewed as universal computers, we have shown *in practice* that the general chaotic properties of nonlinear dynamical systems can perform a variety of computations."[12]

Furthermore, neural networks (and their other computational analogs) demonstrate *superior* information-processing and memory capacity when they attain chaotic dynamics.[13] One would suspect, therefore, that the more densely iterated a processing system is—which is to say, the higher its fractal dimensionality (the larger the number of self-similar scales, hence, the more difficult to distinguish from randomness)—the more complex the processing tasks it would be theoretically capable of handling.

Indirect evidence for this hypothesis can be found in the human cortex itself. A number of researchers have correlated the fractal dimensionality of both EEG and MEG (magnetoencephalogram) tracings with various types of mental activities. They found that fractal dimensionality increases with intelligence and with age. Furthermore, it is high in intelligent people even when the brain is unstimulated; while in others dimensionality attains almost the same height but only during periods of stimulation. Dimensionality is in general increased during imagination of objects compared to perception of them, during creative activity by contrast to deductive logic, and during states of positive rather than negative emotion.

Alas, they also found that people who are in love have EEGs with markedly *reduced* fractal dimensionality (*"Even a god, falling in love, could not be wise."*—Publius Syrus, *Sententiae*, ca. 50 B.C.), along with people listening to pop who prefer it to classical or other complex forms of music.[14]

Chaos and high fractal dimensionality can also readily be found in phenomena that arise out of the reciprocal interactions of human beings as processing elements. The most readily accessible instance is in markets, since the numerical quantification of signals mutually exchanged and reciprocally modified—coupled chaotic oscillators—has already been done for us: price.[15] Especially with the appearance of the Internet, similar models have been developed with respect to global information processing and, indeed, to society itself: "Global Brains as Paradigm for a Complex Adaptive World" was the title of the keynote address for a recent university conference on complex systems.[16]

Furthermore, Barbara Drossel of the Theoretical Physics Group at the University of Manchester has shown mathematically how interacting elements must form groups up to a given size, after which further self-organization causes groups of groups to form up to a given size, and so on. The result is a process that naturally forms nested levels of scale:

> During the past years, physicists have begun to study complex systems like evolution, ecological systems, human systems and economics. The models . . . are usually composed of units that interact . . . and produce a complex large scale behavior. . . . [M]odels for evolution can give rise to [the] . . . distribution of extinction events, . . . models for ecological webs generate several . . . layers of species, models for urban development pro-duce . . . the distribution of cities, and models for stock exchange and company growth show the scaling behavior characteristic of those sys-tems.
>
> . . . [O]ne important characteristic of complex organisms such as life on earth or human civilization is that they have interactions between units of various sizes. Thus a biotope consists of several interacting spe-cies, a species of interacting individuals, and individuals of cells.
>
> . . . If there are limits to the capacity of individuals to communicate with other individuals, the individuals form groups that interact with each other, leading to a complex organism that has interacting units on all scales.[17]

AND ORDER FROM DISORDER SPRUNG: THE QUANTUM VERSION OF CHAOS

We are within a hair's breadth of our conclusion: Very small (even infinitesi-mal) initial differences that arise from *quantum* uncertainty—differences that have no cause—are amplified upward, across all scales, because all scales are iterative and nested. This produces stark bifurcations (or other multiple dis-crete outcomes) at the highest level. The specific outcome that happens in any given case is classical—it violates no mechanical laws—but *which* of many out-comes happens to happen is wholly nondeterministic, in exactly the same way that exactly *where* a given particle strikes in a dual-slit experiment is. We see no quantum miracles in the everyday world that we don't deliberately foster—no superpositions, no vast coherences, no fuzziness. Yet the *particular* mun-dane-appearing reality that we experience is what it is not because of a wholly adequate set of mechanical causes but because of the superpositions, coher-ences, and fuzziness that take place *everywhere* in the universe and at all times and that life has learned to capture at its smallest scale and uses to its great ad-vantage, and which the nervous system amplifies upward because of its com-putational structure.

We're close to saying this, but not quite. So far it all depends on every nested element of a larger-scale parallel-processing system being *chaotic*. This allows the amplification upward. Admittedly this was unexpected—one would have thought all those different chaoses would average out, and they don't. But chaos is still deterministic. So, however complicated, we're still speaking of wholly mechanical systems—no matter how many gajillions of processing elements. If the most fundamental of elements is quantum, hence *genuinely* random, surely it is plausible to assume that such a system would necessarily produce *genuinely* nonpatterned results—that is, pure disorder, not chaos in the technical sense—except where averaging resulted in standard statistical patterns.

For a long time this was in fact believed to be rather obvious. But in the last ten years or so, especially as a result of work published by Steven Tomsovic at Washington State University and Eric Heller at Harvard,[18] yet another new area of physics has dramatically altered our understanding of the classical-quantum interface: The area has come to be known as *quantum chaos*. It is a topic of exceeding complexity and subtlety, and we will be able to touch upon only its highlights. The essential discovery of quantum chaos, first as theory and lately verified experimentally, is this: A simple, noniterative system that "ought" to behave with a very high degree of order and precision is altered by quantum effects to become at least partially random and imprecise, as we've seen in earlier chapters.

But, conversely, quantum effects cause a complex, multiply iterative system of interacting quantum elements to become more *orderly:*

> Quantum mechanics is capable of stabilizing the dynamics of the classically chaotic systems and destabilizing the regular classical dynamics . . .
>
> [Where] the classical mechanics is regular, the quantum wave function may . . . "tunnel" or leak through . . .
>
> [Where] the classical dynamics [would be] chaotic, the quantum evolution . . . eventually exhibits quasi-periodic recurrences to the initial state. . . .[19]

That is, when quantum chaos is present, Poincaré recurrences persist, instead of falling off exponentially or even algebraically. "[I]t takes the flexibility of the quantum wave nature to keep everything from being torn to shreds by the chaos."[20]

A prime example of the first case is the double-slit experiment: The arrival of classical particles at the detector in two narrow bands is fully deterministic; the arrival of quantum particles is widely distributed according to a probability function but random in its specifics.

A prime example of the second case would arise in a system of coupled *quantum* oscillators—for example, in protein folding where multiple quantum entities—electrons, protons—can enter into superpositions and mutually affect each other along with the conformation itself. One would therefore expect, for example, that in the iterative dynamics of protein behavior there should be an unexpectedly large number of Poincaré recurrences—"quasi-periodic" returns to similar states—more often than would happen if the process were classical. This should give protein dynamics a more stably oscillatory character than would otherwise be possible in addition to speeding up its computational tasks:

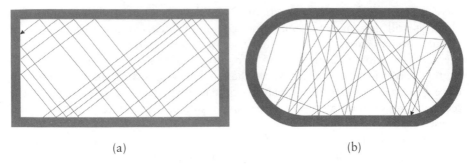

(a)   (b)

FIGURE 16-17   Trajectories of a billiard ball on (a) rectangular and (b) stadium tables.

We show that quantum effects [decrease] the decay rate of Poincaré recurrences in . . . chaotic systems with hierarchical [nested] structure. The exponent p of the algebraic decay $P(t) \propto \dfrac{1}{t^p}$ is shown to have the *universal value p = 1 due to tunneling*. . . . Experimental evidence of such decay should be observable in mesoscopic [intermediate/large] systems. . . .[21]

Here is what this means practically. Suppose that you have an almost frictionless, rectangular billiard table (without pockets) upon which you set a ball in arbitrary motion. It will trace out a trajectory that remains predictable for many bounces before the slight deviations ("initial differences") begin to take a noticeable toll. This system is classical and (practically speaking) non-chaotic, and quite regular. But if you merely round the ends of the table so that it becomes stadium shaped, the trajectory quickly becomes chaotic and therefore much less regular. This difference is illustrated in figure 16-17.

If an analogous experiment is performed with waves, the results are strikingly different. (This has actually been done very recently by Heller and colleagues using microwave photons.[22]) Both the rectangular and the stadium tank generate stable standing wave patterns. Interference reduces the disorder-inducing effects of chaos and allows convergence to a stable attractor, as shown in figure 16-18. Different, discrete ("quantized") energies for the wave/quantum particle yield distinct patterns.

(a)   (b)

FIGURE 16-18   Interference pattern of waves in (a) a rectangular tank and (b) a stadium tank. Both patterns are stable.

In a situation prone to chaos, the wave nature of a single quantum particle induces so much stability that "It is nowadays widely accepted that sensitive dependence on initial conditions does not occur in closed—and generic—single [quantum] particle situations."[23]

In systems composed of *many* interacting quantum elements in at least some degree of coherence, the situation is less straightforward. As it turns out, the combination of quantum plus chaos reintroduces a degree of persistent order to a system that would otherwise quickly become purely chaotic. This order, however, is not perfect, but "fuzzy"—probabilistic—in keeping with all matters quantum.

QUANTUM SCARRING

The kinds of patterns formed by the combination of chaos and interference are like a signature. Once seen, they are recognizable at a glance. They are known as "quantum scarring" of the wave function because of orderly regions (e.g., broad lines) where the probability amplitude is unexpectedly very high. The particle will therefore be found there far more often than seems reasonable—hence it is somewhat re-"localized," more stable, albeit at locations that would be classically unlikely or, indeed, altogether impossible. Figure 16–19, for example, shows another far more complicated arena within with quantum scarring nonetheless appears. It has many characteristics of the stadium wave pattern in figure 16–18b.

The implication is this: A *quantum* particle follows a "trajectory" in which different superposed paths interfere—they add constructively if the amplitudes are of the same sign, destructively if of the opposite sign. Surprisingly, chaotically driven waves of probability amplitude *do* form stable patterns—when they interfere—and these govern the distribution of probabilities. Thus, chaotic quantum systems can actually result in *greater and more long-lasting patterns of regularity* than their nonquantum counterparts.

These unexpected patterns, with their equally unexpected persistence, have no counterpart in nonchaotic classical systems, chaotic classical systems,

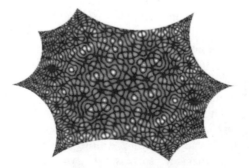

FIGURE 16–19 Quantum scarring creates a high degree of order in what by all rights should be an extremely disordered system. This, too, is a standing wave pattern unique to the energy of the particle(s) and the specific boundary conditions.

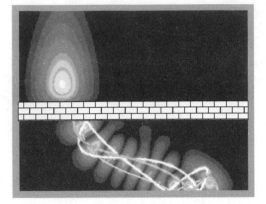

FIGURE 16–20   Scarring due to quantum chaos in a tunneling diode. On the top of the wall (sketched in brick to make the point that it ought to be impenetrable) is the normal probability distribution for an electron—the lighter the region, the denser the probability. Classically, neither a particle nor a wave could penetrate the wall—the region on its far side is "forbidden." The electron can penetrate because quantum tunneling allows it to "teleport" itself there. On the bottom is the "scarred" wave function showing regions where the tunneling electron is likely to be found. The distinct, "organically" repetitive shape of the tunneling amplitude, its intensity, and its persistence very far from the wall facilitate much-farther-than-expected tunneling. The tunneling "paths" of the highest probability (these are the actual "scars") are shown in gray and white.

or nonchaotic quantum systems. They are a unique fingerprint of quantum chaos. Once it was understood on theoretical grounds that such quantum-chaotic patterns should appear, a major experimental drive was launched to find evidence of them, and recently it was found. (See figure 16–20.)

In this instance, the regular ordering of recurrent regions of significant probability amplitude greatly facilitates and extends the tunneling of electrons through and well beyond the barrier. Due to quantum chaos, the tunneling electron will appear at points far more distant than would have been likely without the scarring, and in patterns that are like an organically imprecise variant—a certain degree of, we might say—*coherence*.

There is yet another implication to this. According to the organizers of a recent conference on disordered systems and quantum chaos at the Isaac Newton Institute for Mathematical Sciences at Cambridge University: "Until . . . about a decade ago . . . the conventional wisdom was that all disordered samples are self-averaging, so that [quantum] ensemble fluctuations [among quantum particles] are no more important than, for example, thermal fluctuations in a classical . . . gas. It turns out, however, that . . . [a]*t finite temperatures, quantum coherence sets a new mesoscopic scale.*"[24]

In other words, chaotic dynamics among multiple interacting quantum elements somewhat compensate, surprisingly, for the quantum-destroying effects of decoherence—rather as thermal vibration surprisingly can as well—and similarly *requires* thermal agitation ("finite temperatures"). Aggregate, disordered quantum systems thus can demonstrate orderly quantum behavior (of a unique kind) at a "mesoscopic" scale—midway between the typically ul-tramicroscopic scale of quantum events and the everyday—without isolation from the environment and without coherence. The quantum behavior of its much smaller scale constituent elements is thus reflected in the behavior of the aggregate, rather than being averaged away. The same heat that destroys quan-tum coherence enhances the alternate form of orderliness that results from quantum chaos.

The actual "shape" of quantum chaos–induced large-scale orderliness is not the same, however, as the "shape" of coherence-induced large-scale order-liness. Quantum scarring demonstrates a stable interferencelike patterning, but it "looks" distinct. Only coincidentally, perhaps, it has a more "organic" feel. The shapes generated by quantum chaos in figure 16–20, for example, look less like a mathematically regular set of waves than like the irregular regulari-ties of a vertebra.

There is yet another important feature of quantum aggregates. If the ag-gregate behaves like a spin glass, with truly quantum elements, then clusters of its elements will self-organize to form larger-scale subunits midway between the scale of an individual element and the scale of the whole. These subunits, referred to as "spin glass shards," are likewise larger-scale manifestations of lower-scale quantum behavior. They are absent if the individual elements of the spin glass are not truly quantum.

THE WORLD IN A GRAIN OF SAND

All of the above mechanisms conspire, as it were, to make it plausible that the consequences of quantum behavior at an ultramicroscopic scale are amplified upward to everyday scales.

First, chaos in a quantum aggregate directly pushes small differences up-ward in scale (rather than destroying them, as had been assumed) and makes them more persistent. The presence of tunneling makes quantum chaos more likely, and quantum chaos tends to enhance tunneling.

Second, if the quantum aggregate itself has spin glass characteristics (as do proteins in their folding process), then subregions of the spin glass will form orderly, persistent patterns unique to quantum systems and absent in classical ones. Both these magnified quantum effects are then met by and shape the dy-namics of the next, classical, level. If this next level of organization is adaptive, and thus alters its parameters in response to external pressures (from yet higher-scale influences), then it will influence the lower, quantum level—down-ward, as it were. Intense exercise will increase the demand for certain kinds of proteins. As the body produces more and more of them, and requires more and more of the specific amino acids they require, the environment within which newly synthesized proteins fold will be somewhat altered. The "boundary con-ditions" of the spin glass problem will change and so, too, will the exact nature of the quantum effects and the resulting orderliness. If nature is true to form, one should anticipate that under conditions of large-scale demand, the net changes in the boundary conditions will result in even greater efficiency of pro-tein production and folding.

In short, lowest-scale quantum effects influence the initial state of the next scale; adaptive pressures felt by that next scale shape the boundary conditions of the lowest scale, and hence the exact form of the quantum influences, in a mutually adaptive feedback loop.

Third, if at the next level the system has spin glass dynamics, its yet higher-scale outcomes will show *classical* chaos. This means that its final states (energy minima) will tend to be sharply distinct (e.g., dichotomous) but extremely sensitive to differences in its initial state. Its initial states are sharply discrete due to the quantum chaos that created them.

Fourth, spin glass structure and chaotic dynamics are found at every subsequent level upward, as well, the sharply distinct outcomes of a whole at one level serving as the distinct initial states of the elements at the next level.

Fifth, whether we view a next higher level in terms of their spin glass/neural network/cellular automaton analogies or in terms of coupled iterators, whether stable or chaotic, systems such as these spontaneously self-organize into a hierarchy of scales and compute.

In sum, then, quantum dynamics at the foundational level of a nested hierarchy of massively parallel computers should alter the computational processes of living matter in two ways: First, quantum dynamics greatly speeds up the computation at those levels for which quantum effects are directly present. In the case of proteins—living spin glasses that employ tunneling to achieve their solutions—the quantum properties are intrinsic, classically impossible, and almost certainly necessary for life to be possible at all. (Without genetically determined variation in the amino acid sequence that allows its alcohol dehydrogenase protein to utilize proton tunneling at high temperatures, the bacteria that now flourish near hot ocean vents could not exist.) Second, quantum dynamics alters the final outcomes of computation at all levels—not by producing classically impossible solutions *but by having a profound effect on which of many possible solutions are actually selected.*

Here, then, is how it would work in the brain: Tunneling in proteins allows them to find solutions with blazing speed to the tasks with which they are faced—speed that in a classical world would be utterly unattainable. Indeed, some solutions would be altogether inaccessible. The rapid conformational changes that take place in proteins have a computational nature. Such quantum-mediated conformational changes allow cylindrical arrays of tubulin—individual microtubules—and arrays of networked microtubules to function both as signal propagators and as spin glass–type parallel processors. The parallel processing that takes place in microtubules is likewise partially conformational in nature: The solutions computed by microtubules are expressed in part as a change in their shape, which influences the shape, mobility, and actions of whole cells, and also as parts within a cell. Cell reproduction is a most beautiful instance of this capacity.

The computational activities of the microtubule network within neurons have been specialized to facilitate its unique ability to propagate signals as well as to help it establish connections to other neurons. In this fashion, a larger computational network gets established. Here, again, the parallel processing and self-organization involve both transmission of signals and ongoing material alteration of the structure.

Proteins, microtubules, cells, and brains are a convenient way of identifying the most noticeable scale divisions—four nested brains within the fourth brain. But at each of these levels, there are natural subdivisions. Proteins associate with chemicals and participate in chemical reactions that have a computational nature. Thus one might want to focus on a scale between individual proteins and individual microtubules composed of thousands of proteins. Microtubules form networks with related protein arrays (e.g., actin filaments) and with larger individual proteins. Neurons form clusters that behave as single integrated processing units in a way that is easier to identify as such than do single neurons or even canonical cortical microcircuits. The brain itself is composed of networked subunits. Human society is likewise divided into various groupings that define different scales.

THE QUANTUM MOMENT

We live in a world in which everything appears sane, except perhaps ourselves. Billiard balls do not go in any direction they please; they go only where they must, because of forces acting on them that are in turn what they likewise must be. So it is and has been for all time. Various forms of spoon-bending come and go, bending, for a while, those minds that want to be bent. Mechanism triumphs relentlessly over everything in its path. How is it, then, that quantum craziness prevails in the world of the tiny but seems utterly to disappear at the scale of our daily lives?

It is not that the difference in scale between large and small, and the hugeness of the number of interacting small particles at the scale of the large, averages away quantum effects. Life—especially the nervous system, especially the human nervous system, and most especially the human fourth brain—appears to have evolved so as both to take advantage of quantum effects at the scale at which they have a direct influence—the level of biomolecules—and to amplify that influence upward. This influence does not, however, nor scarcely could, mean that the brain is a massive hand-designed quantum computer, as some continue to hope—man as macroscopic quantum creature who can "go in any direction at any speed, forward or backward in time, however he likes"—but it does result in proteins as miniature parallel processors whose capacities are enormously enhanced, making life possible.

There is more. Parallel processing computation of any type and at any scale is based on a common structure—coupled, reciprocally interacting arrays of elements. These elements may in turn themselves be arrays of yet smaller elements. But chaotic dynamics are intrinsic to such structures. And, as we've seen, when a quantum system is chaotic, it has an even greater than expected capacity to self-organize higher-scale order.

Yet, to all appearances, the mechanical model seems to hold relentless sway. The tiny, magical, quantum differences that appear out of nowhere in the universe result in dramatic differences in outcome at the scale of daily life,

from among the various nonmagical possibilities. Our life as it actually happens looks absolutely mechanical, indistinguishable from a world that really is mechanical. And yet it isn't: The world is shot through at every level with the consequences of a dazzling, infinite array of infinite possibilities—a mind-boggling number of quantum events occurring every instant throughout the myriad of proteins composing vast intracellular arrays, orchestrating the never-ending dance of life in billions of networked neurons and trillions of other cells of the brain and body. Somehow, from among the unimaginable number of possible outcomes generated by all these quantum events, via an unknown, unknowable "influence" (free will? If so, whose?), about which the only thing we do know is that it "is" nothing in the material universe, only one outcome is made real, created freshly at the moment it is selected.

Or, perhaps, as David Deutsch insists, the outcomes *all* are present—a stupendous number of parallel universes flashing into existence at every moment—some splitting and branching without end; others coalescing—from among which, by inscrutable means, one is selected to be ours.

And in all the universe, there is nothing of we which we yet know where this astounding process takes place with greater concentration than in the human brain.

17

QUANTUM RIPPLES

*If I get the impression that Nature itself makes the decisive
choice what possibility to realize, where quantum theory
says that more than one outcome is possible, then I am
ascribing personality to Nature, that is, to something that
is always everywhere. Omnipresent eternal personality which
is omnipotent in making the decisions that are left
undetermined by physical law is exactly what in the
language of religion is called God.*
—Frederik Jozef Belinfante (colleague of Wolfgang Pauli,
instrumental in developing the statistical laws of
multiple electron interaction) in John Barrow,
The World Within the World

Having covered much territory with a microscope, here's the big picture, as I
see it, in light of what we've learned so far.

GOD

Among scientists, mentioning God earnestly remains a faux pas. Only the
ironic mode is safe, of which Einstein was undisputed master. Nonetheless,
something is changing. The great English physicist Sir Arthur Eddington once
warned: "It would probably be wiser to nail up over the door of the new quantum theory a notice, 'Structural alterations in progress—No admittance except
on business', and particularly to warn the doorkeeper to keep out prying philosophers."[1] It is too late for that. Schrödinger's famous cat has long since been
let out of the bag, all nine lives in simultaneous superposition for us to marvel
at and speculate about.

But more important, the incredible shock to the deterministic worldview
delivered by the relentless experimental confirmation of quantum mechanical
theory has taken its toll. It is now a world where quantum teleportation, quantum computation, and quantum cryptography are not only being taken seri-

ously, some have already been implemented at practical scales and are the object of intensive commercial research and development. *Science* reported the first development of self-organizing tunneling switches made of individual molecules that promise to speed up computation by as much as 100 billion times.[2] Working quantum cryptographic keys have been distributed over regions large enough to encompass both Washington, D.C., and Langley, Virginia.[3] In an article entitled, "A Schroedinger Cat Superposition State of an Atom," beryllium atoms were made to coexist simultaneously in two places at once. This is the first step on the way to implementing Charles Bennett's teleportation project at IBM.[4]

It is a world in which one can comfortably argue the dynamics of interference among multiple universes both forward and backward in time; can ask seriously, as did Feynman and Wheeler, whether every electron in the universe is the same one, just reappearing through multiple loops in time, and be doubted but not derided; may urge the development of medical imaging devices that work merely by being there, or of computers that no one need switch on—to work, it's good enough they might have been. Well, in that kind of world, it is in some ways less of a stretch to speak of God. I suspect his bad reputation in scientific circles has less to do with theology than with the sociology of culture and the long history of bad and closed-minded uses to which people have put his good name.

Religion deservedly comes in for more criticism in its failures than science does in its, because religion claims for itself the ability to know what's true, whereas science claims for itself only the ability to quantify the probability of a thing being wrong. A genuine truth arrogantly asserted—that is, without so much as a guess at the likelihood of its being false—is a most pernicious kind of falsehood, far worse than a mere mistake, because it alienates people from it.

In any event, a small number of eminent scientists, physicists more than any other, have begun once again to speak of God. But it's a sticky wicket: When quantum cosmologist and explicit atheist Steven Weinberg publicly debated particle physicist and Anglican priest John Polkinghorne on God and religion at a recent scientific gathering, the reports in scientific outlets were for the most part withering; many made it clear that such a debate should never have occurred, certainly not for the public to hear, and most assuredly not in an inner sanctum of scientific rigor. The mere fact of the debate seems to have evoked emotions among scientists similar, I imagine, to those that would have been evoked among devout believers had Charles Darwin been invited to debate the Archbishop of Canterbury during Easter Sunday services just before Communion.

On the merits, Weinberg won. Losing to a mind of that caliber is no disgrace, but Polkinghorne, no scientific slouch himself, was handicapped for a debate of this sort. Wearing a collar, he was implicitly suggesting not merely the possibility of a God but the impossibility of any other god than that which his own church endorses.[5] Doubt, skepticism, uncertainty, error analysis—these are the scientist's humble, invincible weapons in an intellectual arena.

In 1981 the National Academy of Sciences of the United States posted a formal divorce decree between religion and science: "religion and science are separate and *mutually exclusive* [italics added] realms of human thought."[6] I suspect the incompatibility had less to do with an intrinsic conflict between science and God than between competing dogmas. The preceding estrangement had been long and bitter: In 1860 biologist Thomas Huxley said laughingly that "extinguished theologians lie about the cradle of every science, as strangled snakes beside that of Hercules."[7] Following the divorce, the rancor hardly abated: Joshua Lederberg, Nobel Prize–winning molecular biologist at Rockefeller University, recently noted wryly that "[t]he space available for God appears to be shrinking." Richard Dawkins is more blunt: Anyone who asks "the why question" is simply "scientifically illiterate."

Among the illiterates:

■ Francis Collins, geneticist, codirector of the team that found the gene for cystic fibrosis, currently director of the National Human Genome Research Institute at the National Institute of Health and a devout Evangelical Christian (the kind routinely tagged by the popular press with the epithet "fundamentalist"). "I am unaware of any irreconcilable conflict between scientific knowledge about evolution and the idea of a creator God. Why couldn't God have used the mechanism of evolution to create?" Indeed. Perhaps it is he who all along has tweaked the choices at every quantum branch point—consider how many—to create brains capable of so effectively arguing his nonexistence.

■ In 1964 physicist Charles Townes shared the Nobel Prize for the invention of the laser. He is a former provost of the Massachusetts Institute of Technology. "It is not uncommon for good scientists to be believers," he states. God was a "source of strength" at the time of his historic work, helping him to overcome the crises of confidence that accompany tough challenges.[8] "To get around . . . invoking God may force you to extreme speculation about there being billions of universes. [This] strikes me as much more freewheeling than any of the church's claims." In 1981 "Townes chaired the commission that persuaded President Reagan not to field large numbers of the highly destructive MX missile. Townes says that before each commission meeting, he prayed for guidance."[9]

■ A former dean of Cambridge University, Arthur Peacocke is a biochemist who left research to become a minister. He is warden of the 3,000-member Society of Ordained Scientists.

■ A research study published in *Nature* found that about 60 percent of working physicists and biologists hold strong spiritual beliefs—a higher percentage than among other scientists.[10] On the other hand, among "greater" scientists working today (as the researchers defined these), only 7 percent consider themselves "believers." The highest percentage was among mathematicians—14 percent.[11]

■ David Scott is a former physicist who is now chancellor of the University of Massachusetts at Amherst. "In postmodern academic culture, the majority of scientists think that to be taken seriously they must scoff at faith," he states in an interview. "Yet the truly great scientists were not afraid to ponder larger religious aspects of their work."[12]

■   Christian de Duve, a molecular biologist at the University of Louvain in Belgium and winner of the 1974 Nobel Prize, says: "Many of my scientist friends are violently atheist, but there is no sense in which atheism is enforced or established by science. Disbelief is just one of many possible personal views."[13]

A big problem is "creationism." It is widely and mistakenly assumed that there are only two possible positions: Either one believes in God and that he created Earth and everything on it about 6,000 years ago over a six-day period; or one believes that the world is an utterly meaningless, mechanical machine. "Creationism is an incredible pain in the neck, neither honest nor useful, and the people who advocate it have no idea how much damage they are doing to the credibility of belief"—in the words of a climatologist who also writes on religion.[14]

Francis Collins voices a similar concern: "[B]ecause of the creationists, the standard assumption is that anyone who has faith has gone soft in the head. When scientists like me admit they are believers, the reaction from colleagues is, 'How did this guy get tenure?'"[15]

The reality may be considerably more subtle. Physicist Father Polkinghorne, now president of Queens College at Cambridge University, observed: "The trend is to look for God in dramatic discontinuities in physics or biology, and if none are found, to declare religion vanquished."[16] The director of the Palo Alto Institute of Molecular Medicine agrees: "Just as people came to understand that God does not cause lightning, gradually society will understand that consciousness and other things attributed to the almighty arise naturally, too."[17] The problem is that "naturally" now includes events that have no physical cause.

In other words, ". . . God may act in subtle ways that are hidden from physical science," continues Polkinghorne.[18] Mystics have always speculated about his "hidden" nature: He is there all right, they conclude, but for our benefit, he maintains plausible deniability. "The fool says in his heart, 'There is no God,'" say the Hebrew Scriptures. The mind can reasonably deny him. It is the heart that can be fooled—belief or disbelief both equally capable of being the fool's conviction.

Some scientists foresee a rapprochement of science and religion: "Physicists are running into stone walls of things that seem to reflect intelligence at work in natural law . . . [and] biologists will [likewise] hit stone walls if they fail to find explanations for essential effects like sudden jumps in neurological sophistication. . . . [T]he more we know about the cosmos and evolutionary biology, the more they seem inexplicable without some aspect of [intelligent] design," says Charles Townes.[19] Others insist on just the opposite: According to Richard Dawkins, "The universe we observe has precisely the properties we should expect if there is, at bottom, no design, no purpose, no evil and no good, nothing but blind, pitiless indifference."[20]

In my view, neither position has it right. It is rather a matter of preference—of faith, if you wish—as to how you explain the quantum foundational basis of the universe, its concentration in life, and its amplification by the brain: Either it is absolute chance or absolute will. Opposite as they sound at first, between the two science can point to no distinction and the universe, it seems, will offer no evidence. Both are equally mysterious as explanations go. Indeed, they are hardly even that: They are merely terms for something beyond our ken. You might as well call it the Tao—or Ralph. In the words of theoretical chemist Michael Kellman, "In a world where 'choices' appear to be constantly being made between different chance outcomes, the idea of divine intervention may not seem quite so absurd—or at least not so much more absurd than the bizarre things we already know about the world."[21]

Furthermore, if you decide that will is the best name for it, then from neither science nor from nature may you may expect a clue as to *whose* will it is.

The one thing you do have to reckon with, however—or should admit that you have to—is that there is *something* going on, everywhere, that creates the particular world in which we live, a creation that occurs not just once, at the beginning, for all time, but always, just as moment to moment it sustains who we actually are in that world. To my mind, it is as big a misreading to claim that science tells us that this something cannot be God as to assert that science tells us it must be. The world may be pregnant with hope and meaning and purpose beyond our brief, selfish lives, or it may be meaningless—random to an previously unimagined degree. But one thing it is not: mechanical. Perhaps, as many ancient Gnostics sincerely believed, the world is a gigantic practical joke implemented by a second-tier minor deity out of boredom. Nor is my writing this a joke: One eminent physicist, Allan Harkavy from the State University of New York, recently published an essay—safely, as he is now emeritus and can say what he pleases—entitled "Speculations Concerning Will and a Local God."[22]

FREE WILL

The majority of scientists, especially those who study the brain, consider the question of whose "will" not worthy of asking, let alone answering. Nonetheless, the fact that the entire operation of the human brain is underpinned by quantum uncertainty, and the likelihood that the brain is structured so as to make actual, large-scale outcomes mechanically indeterminate, casts a new light on the question. Neuroscientists may certainly continue to elucidate with ever greater precision the mechanisms by which small-scale operations are transferred from one scale to the next without ever making appeal to quantum uncertainty and without ever needing to. But, in the end, the answer to the question "Why did I do this and not that?" will remain forever unknowable.

Among the possible answers: "Nature made me do it"; "God made me do it"; "the Devil made me do it"; "my upbringing"; "my economic status"; "the Tao"; "Not I, but Christ in me"; "I had to" (a favorite); "them" (another favorite); and of course, simply, "Because, on balance, I wanted to." Mysticisms of all sort have long made the dangerous point that, in the end, there is a profound difficulty in distinguishing between God's will and man's. If one wants to, one might say that the only essence of God of which we have knowledge is his seemingly infinite capacity to choose—selecting from among quantum alternatives the one that shall be. If the human brain is designed so as best to capture, distill, and concentrate that essence—at least from within a very small region of the universe—perhaps our will is in some sense a miniature portion of his. Or perhaps our will is nothing more than an unusually dense distillation of randomness. In any event, it is no less free than the universe itself, and possibly considerably more. Beyond that, science won't say because it can't.

CONSCIOUSNESS

Stuart Hameroff's fascination with microtubules came naturally to him. As a practicing anesthesiologist as well as an academic researcher, getting rid of consciousness and restoring it is his business. For a curious mind it must be irritating to be so in the dark about the object of one's sophisticated, daily, life-saving manipulations. Roger Penrose's two books about the mind and brain were likewise chiefly concerned with explaining consciousness. By definition, consciousness is the most ubiquitous of all human experiences, but as Zen Buddhism has long taught through jokes and riddles, the most impenetrable. I think it will remain so, in spite of all the attention "consciousness studies" now attract.

In the 1960s, the hope of finding a quantum foundation to life was expanded to encompass consciousness. The early quantum pioneers suspected there might be such a connection. Someone who makes a quantum measurement seemed to become part of a whole that now unfolds in coordinated fashion according to quantum laws. Indeed, someone who merely *intends* to make such a measurement appeared to have like effect. No wonder people began wondering whether there had to be some intrinsic connection between mind and matter at the quantum level, since such "influences" are absent in the classical world.

On the other hand, we now know that accidental "measurement" has the same effect—say a double-slit kind of situation just happens to get established, between or within molecules, for instance. (This actually happens when water dissociates.) If a certain kind of quantum computer can solve a problem whether it's turned on or not, merely because it could if it were, the intent becomes irrelevant. *What counts is the physical arrangement of stuff.* What happened to consciousness? No one really knows.

I think that the keenest observation pertinent to a relationship between quantum mechanics and consciousness was made by William James, the pragmatic father of American psychology, before quantum theory was developed. Any mechanistic neuroscience, he noted, must make of consciousness a passive bystander at most. In such a view, the very "molecules," as he called them, that make up the moving parts of the brain are moved in the only way they can because of the mechanical (or electromagnetic) effects of other molecules upon them. They, in turn, transmit these effects to others. In the end, no motion of brain or body can be or could have been anything other.

But this means that the mind associated with the brain can have no effect whatsoever—it cannot affect the motions of even so much as a single molecule, since all motions are already wholly determined. Free will is therefore a complete illusion. Since the choices executed by free will would be the only measurable indications of consciousness, then if free will is at best an illusion, so consciousness may be as well.

Consider: Suppose one could construct a Golem, with processing characteristics identical to those of human beings but lacking consciousness. Suppose further that the "man is a machine" hypothesis is correct. Then operationally, between the behavior of this android and that of a man there would be absolutely no distinguishing. It would walk, talk, and work, just like a man. It would speak of itself in the first person, express joy and dismay and love, express sincere remorse, appear to us to take cruel pleasure in the sufferings of others, seek salvation, perform scientific experiments, discourse with learning about all matter of things, hang its head in what looks like embarrassment at its gaffes, fight wars, defend its pride, cry for justice, and philosophize deeply about its own nature *and consciousness* all because mechanically, it can do nothing else.

Most neuroscientists adopt the position expressed before that "consciousness and other things attributed to the almighty arise naturally, too." But this is a subtler statement, more hedged, than it may seem. To say that something *arises* out of natural processes is not to say that it is *a logical consequence of natural processes.*

Here's the distinction. Everything in this book is congruent with the hypothesis that consciousness simply emerges as natural processes unfold. These natural processes include, as it happens, quantum phenomena, simply because they are a feature of the physical world—if not precisely as I've claimed, then in some way. At what scale they are a feature is besides the point. (Remember Bob and his new bride.) The physical world is whatever it is, and as a result of processes natural to that world, life seems to have emerged from them and, via its natural capacity to compute in parallel, tends over time to self-organize at ever increasing densities of computational iteration.

At some point in this process—who knows when?—the phenomenon we call consciousness, or the illusion of a phenomenon called consciousness, or the capacity to have illusions makes its appearance. If a God created the world and intervenes in it everywhere at all times, then the appearance of consciousness via his first and all subsequent interventions (which is how one might prefer to think of the quantum selection process) is as much his doing as are all things in the universe. If we'd like, we may say, in brief, "His will created our wills." (You might prefer to say her will, or theirs. Science cannot gainsay your preference.)

On the other hand, if no such God did any such thing—or more simply, if that's how you prefer to see it—then it still remains a fact that consciousness emerged in association with natural processes, evidently related to complexities of self-organization; and it's possible to understand, along the lines of what's been laid out in this book, the unexpected and important role of chance in those processes—both the stochastic kind of chance at scales where quantum effects play no role and the absolute kind of chance at scales where they do.

Most modern neuroscientists go further, however. They say, "One day we will understand *how* the organization of the brain gives rise to consciousness," meaning how such and such an organization of matter *necessarily must be conscious.* This is a lot more than saying that we have *found* such and such an organization invariably to be associated with consciousness. I think it very likely that we will learn to create machines that are conscious; I think it equally likely we will never understand *why* they're conscious.

I have studied a good deal of the newer writings on consciousness and neuroscience as well as those on consciousness and physics and on consciousness and philosophy. In the end, I have thrown up my hands. Perhaps it is my own limitations, of course, but here's what I've concluded: I doubt we will ever be able to show that consciousness *is a logically necessary accompaniment to any material process,* however complex. The most that we can ever hope to show is that, empirically, processes of a certain kind and complexity appear to have it. Perhaps it is an intrinsic "quality" of matter, like mass. Maybe it's somehow related to the foundational nature of "information." In any event, I have found almost all the writings on this topic singularly confused, filled with the wishful biases of the writers' professions.

The most confused of these claims—very fashionable—is to assert that like all biological processes, consciousness has evolved via natural selection because of the survival advantage it confers. Surface plausibility notwithstanding, the notion is confused. First, note that it is put forth most forcibly by the purest mechanists both to deprive mystics of a favored domain of mystery (Master: "What is the sound of one hand clapping?" Disciple: "Forty Hz oscillations in the parietal cortex are self-generated when conflicting signals from the reticular activating system and . . .") and because it is an obvious challenge to any neuroscience that hopes to claim a reasonable degree of completeness. So, for example, Patricia Churchland writes, "[I]t appears highly probable that psychological processes are in fact processes of the physical brain, not . . . of a nonphysical soul or mind. . . . Materialism is the more probable working hypothesis."[23]

Now, Churchland et al. may be correct. But it is precisely in such a purely mechanical, materialistic model of the brain that there can be no place for a mind that exerts influences on the matter of the brain—because, as Churchland says, in that view, there's no such mind there.

Moreover, the materialist hypothesis is inextricably linked to Darwinism. A fuller citation of Churchland's first phrase above reads: "Despite the rather remote possibility that new discoveries will vindicate Descartes* [e.g., the reality

---

* Reader, caution: This is a misleading reading of Descartes, whose splitting apart of body and soul was the crucial step that led *to* materialism, not away from it, a consequence Descartes neither foresaw nor intended. His claim was that there may be a soul (and spirit) but these can interact with matter only by divine dispensation—an exception to natural law. From a law-abiding, hence inefficacious God (or will) to none is not even a small step, it's a hiccup. With friends like Descartes, however well-intended, God needs no enemies.

of a "nonphysical soul or mind"], materialism, like Darwinian evolution, is the more probable working hypothesis." The Darwinian claim as applied to consciousness says that it evolved because it was beneficial to adaptation. How can something that in the materialist view can't exist be beneficial to adaptation?

Is it rather the *illusion* of consciousness that is adaptive? But how can a machine without a mind have "illusions"?

Perhaps, then, there is such a thing as "mind," except that it can have no influence whatever on matter, because materialism requires no nonmaterial influences. A nonefficacious nonmaterial presence (which would also be undetectable, since detection requires interaction, at least outside of quantum effects), however pointless, is, I suppose, a logically acceptable alternative. But how can something that has no effect enhance adaptation?

It is reasonable to argue, as Churchland also has, that simply because we don't *now* understand how something happens does not mean that we *never* can understand it. But the mysteriousness of consciousness, it seems to me, is self-evidently of a different order from other puzzles. Unlike lightning, say, consciousness cannot even be shown to exist. Indeed, along the lines of the Golem example, there are plenty of good arguments that it doesn't.

No, let me put it more strongly: The arguments that consciousness doesn't exist are vastly superior in their consistency and logic to any argument that it does, let alone that it necessarily arises from the interactions of matter or has evolved while yet having no reality of effect. Belief in consciousness is just that: a belief. And beliefs are themselves products of minds. So if there's no mind, there are no beliefs and no illusions that there are minds with beliefs.

Out of this kind of infinite regress I have seen no one offer a plausible hope for escape. Furthermore, even those who accept the logical consequences of their own premise and assert that there exists no will, no self, no ego, no mind—the radical behaviorists come to mind, such as B. F. Skinner—invariably make their assertions willfully, defend selfishly their (usually) academic turf, display all the signs of wounded ego when criticized, and take evident pride in the quickness of their minds. Intellectually, there's nothing wrong with asserting the materialist hypothesis. Perhaps it really is the truth. But then, as for consciousness, the only consistent statement should be "Beats me . . . and so far as I can see, it bids fair to keep on beating me."

MORALITY

Of course, the perfectly logical defense that could be offered by the willful, selfish, egotistical intellectual who won't believe in will, self, ego, or mind (by contrast to the willful, selfish, egotistical intellectuals such as myself who do) is that they can't possibly act in any other way (and neither can I, they would have to concede—but I don't). But, look, here's a surprise. We share a common attitude. (Hmmm. Can mindless brains have attitudes? Whatever.) We both "believe" that to be "too" willful, selfish, egotistical, and clever is "wrong."

That's pretty odd, when you think about it. If everything is as it must be, period, then what possible meaning could there be to the notion of something

being "too" anything, or "wrong" or "right." Like or dislike what you will . . .
no, that's not right: Like or dislike what you must—your preferences are unreal. Sure, go ahead, march for nuclear disarmament, vote Republican, care for
starving people in Rwanda, protest abortion, protest abortion protestors, pro-
test abortion protestors protesting abortion protestors—what else can you do?
Kill the Jews, put on trial people who kill Jews, put on trial people who put on
trial people who kill Jews. It's all the same: ridiculous to call one "right" and
the other "wrong." Might makes right. That's not a philosophy—it's an obvious fact, if there's no free will.

So what if it's what the Nazis taught—it's being taught now at Harvard
Law School: Law isn't based on morality, it's a technique that oppressors use to
keep the oppressed in their place. You don't like the fact that the argument can
be equally and enthusiastically embraced by oppressors? Of course you don't
like it—you can't help that, just as the guy over there can't help feeling thrilled
and liberated by it. But so what?

You disagree. There are certain things, you insist, that are simply good
and right, and others that are bad and evil. Of course say you—as you must.
But I must say that's not so. That some of us are forced to dress up our actions
in the language of free will and moral choice says nothing about whether either
exists. And if there's no choice, there's no moral choice and morality is a meaningless word.

In brief: It has become widely believed that "man is a machine," especially
among intellectuals. The consequences lie all about us. That doesn't mean it's
not so. But because the consequences seem to me so bad, I indeed hope it isn't
so. Quantum theory may once again make reasonable an insistence that we are
free; it offers no guidance whatsoever as how to use that freedom.

RELIGION AND BELIEF

It seems to me evident that some set of rules, and the beliefs that sustain them,
are the best because they are the most real. I do not believe that all or even
many different sets of rules are equally valid, because they can't all be equally
true. I think it quite possible that one belief system may have gotten it most
right—so much the worse for the pride of those who are less correct.

I have my own guesses as to what these beliefs are—as do we all. I may
very well be wrong. Indeed, I've changed my mind about things so often that
I'm positive I am wrong in many ways as yet unknown to me—how can one
learn anything without first being ignorant on the subject, or wrong? So, without falling into the mistaken notion that all belief systems are equally valid, I
respect the right of others (their obligation, in fact) to adhere to the beliefs they
consider most correct. Many of them, I realize, will be more wrong than I,
many less wrong. In any case, it is humiliating to picture myself ridiculing them
for their convictions, no matter how wrong I think they are, as has now become the norm in "civil" discourse, especially by those who most loudly
preach "tolerance."

The one camp I find deserving of fierce opposition is the one that allows
for itself no serious possibility that it might be wrong. In this camp may be

found as many materialists as religious "fundamentalists." Their hallmark is the pleasure they take in clever ridicule.

THE FUTURE

What the quantum foundation of life tells me is this: The future has not yet been written. In very large part, with dramatically opposite possibilities lying as yet untapped, we will make it what it will be. In a strange and eerily perfect way, we are condensers of destiny, distilling the maximum possible amount of the universe's intrinsic freedom into an incredibly tiny package. I don't know whose freedom it really is, if anyone's, but I am convinced it is there and that it is woven into every fiber—every tubule, I suppose I should say—of my being, even if one calls it "chance." Perhaps if I choose to flap my finger here, a tornado will erupt in Texas, or joy will break out. (So it behooves me to do the best I can to learn whatever rules of conduct are most likely to yield the results I long for.)

I don't know for sure whether there are such things as "prophets" in the biblical sense. But if there are, it seems to me that they, too, don't know the future so much as the possibilities that lie before us. "The Messiah will come in an age that is either entirely good or entirely evil," says an ancient Jewish legend from long before the days of chaos theory and bifurcation. Perhaps he hasn't yet made up his mind either.

But the following does seem likely to me. For precisely the reason that consciousness is in some mysterious way associated with a certain kind of organization of matter—a densely iterated, nested hierarchy of matter engaged in parallel computation—we are one day soon very likely to create conscious devices, for good or for evil.

Perhaps this is the next step in evolution, or in someone's plan: to create mind required the extraordinarily long, slow process of self-organization via natural selection—4 billion years. But with the (seeming?) appearance of mind, the distillation process achieved an intensity such that the next steps will be deliberate: a careful adaptation of the best that evolutionary computation produced to speed up the process a trillionfold. I think it quite possible that, in my lifetime, we will have at least come close to creating devices whose intelligence and capacity to self-propagate and evolve even further will greatly outstrip our own.

Perhaps the first appearance of truly self-evolving machines will mark the end of our own evolution. In any event, I think it far more likely that we will either exceed or destroy ourselves than that we will understand ourselves. It is no longer just Hugo de Garis in Brussels who is setting out to develop self-evolving hardware brains. Both NASA and the German research branch of IBM are doing likewise. I would imagine that Microsoft, a company the market value of which now exceeds half a trillion dollars (only nine nations worldwide have larger gross national products, Spain ranking just above Microsoft), is quietly but fiercely doing the same. (Its research department recently hired one of the world's most eminent mathematicians to work on quantum computation algorithms.)

Self-evolving brains composed of teleporting quantum computational elements processing information simultaneously in multiple universes? I don't think the future is remotely imaginable, since this is all just beginning. Or is it just science fiction? It's not. Between July 16 and 19, 1999, at the Jet Propulsion Laboratory, NASA and the Department of Defense held their first annual conference on evolvable hardware. A sample of the presentations:

- Evolving Circuits by Means of Natural Selection
- Embryologic Electronics
- The Design and Use of Massively Parallel Fine-grained Self-Reconfigurable Infinitely Scalable Architecture
- Self-Repairing Evolvable Hardware
- Genetically Engineered Nanoelectronics
- Co-Evolutionary Robotics
- Evolving Wire Antennas

SOCIETY

Pretty much throughout history, the barriers that separated haves from have-nots were only slightly permeable—race, nationality, ancestry, mores, habits, style could be changed only somewhat or not at all. But there have always been individuals who vaulted these barriers to raise themselves and sometimes their fellows, as well. Occasionally whole groups would ascend from slavery to conquest: The ancient Jews did it; so did the late rulers of the Roman Empire and the French citizenry at the time of the Revolution and Terror. Wholesale class transformations were rare in ancient times, far more common lately as society has grown dramatically more meritocratic.

But the new class barriers that I now see looming are likely to prove both far more fair and far more cruel than any that have preceded them. We are in transition, between an era that has died and one waiting to be born. The hereditary aristocracies have vanished; the new cognitive elite has just begun its ascent to dominance. For now, for a while, the barriers are at their lowest. The Silicon Valley phenomenon is but the hors d'oeuvre course for the banquet to come. To this banquet everyone will be invited—race, background, class, manners: None of this will matter much at all. But how many will be intelligent enough to understand and accept the invitation?

Very few. Math and physics students are now being recruited by Wall Street because even finance is becoming ever more quantitative. The right* shoes and the right* social network may still help at the beginning, but increasingly it's the right number on your Stanford Binet, and the application of that number to the right stuff, that does the trick. It's not for nothing that the riches of Silicon Valley were created by nerds, that so many of the outstanding computer and neuroscientists started out in physics and math.

* white

To get onto this track, the best students at the very best schools begin advanced studies in mathematics and modern science at an ever younger age. Andover, for example, offers a mathematically rigorous class in special relativity to its high school–level students. But at the same time, the general quality of education (in America, at least) has declined relentlessly. The majority of students are entering college ever more poorly prepared for quantitative studies. Many students fall quickly away, and spend their time—for obvious reasons—studying and concocting philosophies of phony egalitarianism. The result of these opposing forces is that a small elite of the very best is getting farther and farther ahead of the pack and is now creating the future. Of all the major universities, MIT now produces the largest proportion of entrepreneurs and the largest number of new businesses per capita per graduating class.

When members of the rising elite have consolidated their ability to manipulate the emerging biomolecular and quantum computational technologies, they will form a club whose barriers to entry will be the most scrupulously fair in history—and the most ruthlessly impenetrable to the unqualified. Having little need to preserve dominance by force or trickery, they may form, if they have a mind to, the most benign ruling class imaginable. Or maybe not. Either way they will be envied and feared. If they are ever overthrown, it is more likely to be parricide than fratricide: Their own creations will have the capacity to do the deed; their subordinate brethren won't have a clue.

Perhaps we will even alter human nature itself and turn ourselves into something utterly alien, a race for whom the old standards—wisdom, humility, nobility, kindness—will be discarded like a serpent's skin. We'll just become winners, until we meet a race that's better at it than we.

It is, in other words, the best of times and the worst of times, a Janus superposition of possibilities about to reveal one of its two faces. Willy-nilly, we have embarked on an adventure that leads to shores far more distant and alien than any we have ever set out for before. I believe that if, as we reach for the sky, we do not at the same time root ourselves in the wisdom that gave birth to our culture, we will destroy ourselves.

But of all the possibilities, however remote, I rather like this one. Perhaps we will create a new race of creatures, not out of ourselves, but set into motion by us, who will strive for, reach, and maintain our old standards in a way that we repeatedly glimpse, but rarely embody, and most often reject anyway. In their kindness, they will be willing to keep us—flawed racehorses turned out to pasture, kept from the cannery out of a gentility that we never could attain. It's likely that such creatures will have discovered the large answers that have eluded us for so long. I picture myself asking them, eager to know the new answers, the real answers, at last. I listen intently, like a parent with Alzheimer's disease hanging on every word of his now-grown child as he patiently explains something rudimentary. The child looks down in concern as he answers, choosing his words with care to ensure that my feeble mind can grasp it. He finishes, smiling tenderly. But I fall back, puzzled and frustrated. There was nothing new. I had heard it all before. I still don't get it.

Appendix A

THE SIMPLEST HOPFIELD NET AND THE "CANONICAL CORTICAL MICROCIRCUIT"

Figure A–1a shows a network of three neurons (spins), each of which can take on the values of either +1 or –1. They are reciprocally connected with mutually identical connection strengths (coupling) and every neuron is connected (every spin is coupled) to every other neuron (spin).

Memorization and recall mean that the network is to be presented with a set of patterns to learn. After it has learned (stored, or memorized) them, it will be presented with partial or distorted versions of the originals. If it has memorized the patterns properly, these cues will trigger a series of "associations" that will converge on the correct original. The network will then settle in the configuration evoked by the stimulus and will emit, as output, the correct pattern. Think of the game "Name that Tune," if you will.

Now, such a minute network can hardly be expected to be able to memorize much. There are, after all, only eight possible states in which the network can exist, given that its neurons can each take on only two values. The eight states for the three neurons are: (–1,–1,–1),(–1,–1,+1),(–1,+1,–1),(–1,+1,+1),(+1,–1,–1),(+1,–1,+1),(+1,+1,–1),(+1,+1,+1). Of these eight, let us select two to be learned as "fundamental memories" (valleys, in the energy landscape): (+1,–1,+1) and (–1,+1,–1).

If we present these two patterns to the network and allow the connection strengths between neurons to be affected in accord with Hebb's learning rule, the connections will settle into the configuration of figure A–1b, with mutually reciprocal weights of either +2/3 or –2/3 (negative connection strengths imply inhibition).

This simple microcircuit has the essential features of a spin glass: reciprocal excitation or inhibition, interconnectedness of elements. In a three-element network, nearest-neighbor connections automatically ensure that every ele-

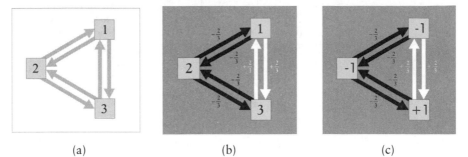

(a)   (b)   (c)

FIGURE A–1   (a) Canonical microcircuit variant as a three-neuron Hopfield network, equivalent to a three-element spin or oscillator network. (b) Canonical microcircuit variant with connection strengths (couplings) that establish frustration, so that it functions as spin glass or as a Hopfield neural network. (c) Initial input state (–1, –1, +1).

ment is connected to every other. In larger systems, to connect every neuron to every other requires many crossing connections. In actual spin networks in nature (i.e., in nonliving physical systems), somewhat wider neighborhoods are created by somewhat longer-distance spin interactions.

Now, let's see what happens when we present the network with distorted versions of what it has memorized. Suppose the inputs to neurons 1, 2, and 3 are such that they get activated initially in the pattern (–1,–1,+1). (In the fourth brain, such input might come either from other similar microcircuits or from lower brains). We'll label each neuron with their initial response so that the network now looks like figure A–1c.

The network takes these activation values and acts on them, updating the state of each neuron and thereby changing the state of the network as a whole. If the net input to a neuron is positive, the neuron goes to (or remains in) state +1; if the net input is negative, the neuron goes to (or remains in) –1; if the net input is 0, the neuron remains unchanged. Table A–1 illustrates how each neuron is affected by its neighbors to generate the subsequent state, given these specific couplings.

Neuron	Net Input	Comment	Result
1	$(-1) \cdot (-\frac{2}{3}) + (+1) \cdot (+\frac{2}{3}) = \frac{4}{3}$	Change +	+1
2	$(+1) \cdot (-\frac{2}{3}) + (-1) \cdot (-\frac{2}{3}) = 0$	No change	–1
3	$(-1) \cdot (-\frac{2}{3}) + (-1) \cdot (+\frac{2}{3}) = 0$	No change	+1

TABLE A–1   The second state of a three-element Hopfield net (spin glass) is generated by the reciprocal influences between each element and its two neighbors.

Input	Output	Basin #	Distance of Input from Basin
[−1 −1 −1]	[−1 +1 −1]	2	1
[−1 −1 +1]	[+1 −1 +1]	1	1
[−1 +1 −1]	[−1 +1 −1]	2	0
[−1 +1 +1]	[−1 +1 −1]	2	1
[+1 −1 −1]	[+1 −1 +1]	1	1
[+1 −1 +1]	[+1 −1 +1]	1	0
[+1 +1 −1]	[−1 +1 −1]	2	1
[+1 +1 +1]	[+1 −1 +1]	1	1

TABLE A.2   All possible inputs and outputs for a three-element network. "Distance of Input from Basin" lists how many of the three input values are different than the three values of the most similar of the two final states.

In other words, *input* (−1,−1,+1) → *output* (+1,−1,+1). But this state is one of the fundamental memories, so it will change no further. There are no intermediate states, as we could have predicted by noting that the initial input differs from one of the two final outputs (one of two basins of attraction) by one "error." That's the one to which it is "nearest" and in which it will settle. In fact, this three-neuron network is so simple that of all possible inputs, six converge in a single step and the other two are the fundamental memories themselves, as shown in table A−2.

In this "toy" example (the simplest possible, in fact), any input is either 1 error away from one of the two fundamental memories (and will go to that memory in a single step), or it is 0 errors away from one of the two fundamental memories (and will stay at that memory). Any successive internal feedback leaves the network unchanged, having converged at one of the two stored memories.

The energy landscape for this network has two basins of attraction each surrounded by three discrete positions and looks something like figure A−2.

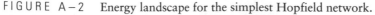

FIGURE A−2   Energy landscape for the simplest Hopfield network.

Each triangular plane within the landscape represents one of the six un-stable higher-energy configurations. Each bottom point toward which the planes funnel the ball represents one of the two low-energy stable states into which the network will settle, depending on where it starts. It's easy to see that a large network composed of a great many of these subunits would function very much like a network of binary elements. In particular, two subunits passing opposite states back and forth will form a two-state oscillator; additional units will establish multistate oscillators.

Appendix B

THE TUSZYNSKI
MICROTUBULE MODULE

The early Hameroff model of microtubule computation assumed no forces affecting the constituent tubulins other than those generated by the tubulins themselves (and these were modeled only very crudely). This meant that the "randomizing" effects of heat (thermal agitation of the tubulins) was not taken into account, as if they were operating at absolute zero, which is biologically unrealistic. In the Tuszynski model, real physiological temperatures are assumed to generate a significant degree of uncoordinated agitation among the tubulin dimers. Because its other features faithfully reflect nature's design, the randomness it incorporates turns out to enhance, rather than diminish, the capacity for intelligent processing in microtubules. This is the same principle as had already been found at other scales in the nervous system.

Now recall the work of John Ross, whose studies of oscillating chemical reactions laid the groundwork for an understanding of cell biochemistry as computation. In 1979 Ross published a particularly nice paper on how oscillating chemical reactions obey the same mathematics as the establishment of domain boundaries in magnetic materials.[1] (Recall, too, that "ferromagnetic" matter establishes many abutting domains, each with parallel magnetic spins. See chapter 6.)

The precise dynamics whereby a domain boundary spreads in a globally ordered pattern corresponds to how a wave of oscillatory chemical reaction spreads, like ripples from a stone dropped in a still pond. The synchronized, traveling global pattern arises out of many purely local interactions. Domain boundaries are marked by competing spin alignment, regions where each spin is subject to both parallel and antiparallel tendencies (or, in cellular automata, regions of simultaneous competition and cooperation), a phenomenon known as "frustration." If frustration is widespread, the substance is called a spin glass.

The equations for such complex systems are exceedingly difficult to solve. In the case of a ferromagnetic domain boundary, the sole solution was discovered in 1972.[2] It looks like a traveling "kink," or "front," that represents the moving domain boundary or chemical wave. (See figure B–1a.)

A traveling, kinklike boundary is therefore a possible mathematical consequence in any array of objects each of which is capable of assuming one of

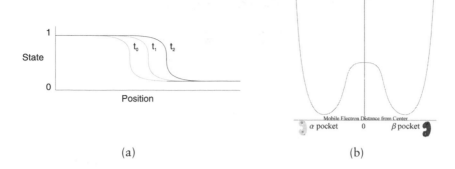

FIGURE B-1   (a) Traveling kink or boundary that emerges from the requirements of interacting ferromagnetic spins and chemical waves. (b) Double potential well for the tubulin dimer. The favored locations for the mobile electron are in a pocket formed by the subunit proteins.

two opposite states and is influenced to choose one by the states of its neighbors. (Whether a wave develops or not depends on the precise strength and direction of these influences.) In 1994 a research group showed that the mechanical forces in microtubules were such as to cause a traveling kink of just the kind that Ross had found in oscillating chemical and ferromagnetic spin systems.[3]

Following up on this finding, Tuszynski made a careful analysis of the exact structure of microtubules., seeking a working, biologically accurate model for MT processing. He took into account the additional fact that each component tubulin molecule carries a "dipole" charge—it has one positive and one negative end, but on balance the charge is zero. Such a dipole "flips" because it has two equally favorable low-energy conformations, depending on which subunit contains a certain "mobile electron." By "mobile" we mean that the electron tunnels abruptly from one location to another. Coincidentally, the tubulin electrical energy landscape as a function of the position of the mobile electron looks a lot like the ammonia energy landscape as a function of its tunneling nitrogen, as illustrated in figure B–1b.

Tuszynski also took into account the precise spatial alignment of a tubulin vis-à-vis its neighbors, an alignment that changes when the dipole charge "flips." Relative to the axis of the microtubule, the axis of alignment of a tubulin is skewed in two different ways, depending on which of its two conformational states it is in. The result is a pretty complicated "neighborhood." Nonetheless, Tuszynski found that when mathematically modeled, the dynamic behavior of a closed, skewed tubular array of these flippable dipoles obeys the same equations as Ross's traveling waves and as ferromagnetic domain boundaries. It forms a traveling "kink," a wave of conformational change propagating along the length of the microtubule. (See figure B–2.)

The solution that satisfied all physiological conditions allowed for traveling conformational kinks to be propagated not only along MTs as a whole but

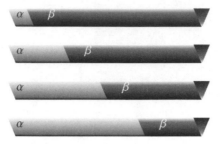

FIGURE B-2   Conformational "kink" traveling along a microtubule. "α" represents a domain where the mobile electron is in the α subunit of the tubulin proteins, orienting the dipole in the direction of travel. "β" represents a domain where the mobile electron is in the β subunit of the tubulin proteins.

along each of the so-called protofilaments, chains of dimers that seem to run lengthwise along an already assembled tubule.[4] And earlier analyses had suggested the possibility of "torsional" (twisting) forces propagating similarly.[5]

Furthermore, Tuszynski found in his model that the presence of an electric field aligned with the tubule would have a significant effect on these traveling waves, either facilitating them or dampening them, depending on the field strength and direction. The presence of such fields is not something that can be taken for granted in living matter—except, as it happens, in neurons. For the electrochemical "action potential" that constitutes a traveling nerve impulse does indeed produce a traveling electric field with a potential effect on MTs in neurons. This field is perpendicular to the lengthwise orientation of MTs in axons and dendrites, so one might assume that it would have no effect. But because of the skew to the tubulins, there is a significant effect. In other words, the adapting pattern of activity of a neuron as it participates in a living neural net alters the signal processing happening in its own internal microtubule network. But the processing characteristics of every individual neuron in the net are likewise affected by changes in the internal microtubule network of each. In short, the neural network of the brain is coupled to the internal neuronal microtubule network and vice versa. The microtubule network therefore adapts to the demands of the brain as a whole, with neuronal plasticity as the intermediary. Different computational scales are thereby coupled.

Another part of the puzzle was put into place a few years after publication of Tuszynski's 1993 paper, by a group of three researchers at the Université de Yaoundé in Cameroon, who found that the physics of tubulin embedded in an array allowed for no stable intermediate conformational states. Each molecule really did "flip," very discretely—quantized (but not quantum!). If the curve in figure B–1a is very steep, its width is narrow and the transmission time from one state to another is short. This implies that "with the numerical values of [Tuszynski's] model . . . kink excitations cannot propagate over long distances whatever the [plausible] value of the electric field [by themselves]. . . . They can move only because of external fluctuations. . . . [T]he motion has an activated character, showing hopping from one [preferred conformational state] to another."[6]

In other words, heat-induced random agitation, associated with orderly electron tunneling, is required for propagation. There would be no boundaries and no traveling waves unless there was just the right amount of thermally induced jiggling. Too little, and the whole thing would "freeze" in place. Too much, and the tubulins would change shape at random—like a piece of iron demagnetized by being heated. Tubulins have adapted to their "ecological niche," taking advantage of the conditions within which they exist, to utilize electron tunneling for the purpose of signaling.

However, if there was merely the right amount of heat, microtubules could carry signals but still wouldn't necessarily compute. They'd be like ferromagnets but not like spin glasses, which are a mixture of ferromagnetic and antiferromagnetic domains with lots of "frustration"—MTs wouldn't be frustrated enough. They'd be intracellular signal connectors, as it were, but not neurons. In genuine spin glasses, frustration is created by physically mixing ferromagnetic and antiferromagnetic particles to create a finely grained "composite." There seems nothing similar in MTs.

But Tuszynski found that there were actually *two* main sources for the needed external activation: one was heat, as mentioned. The other was purely mechanical. In other words, external tension (pushing, pulling, or twisting) of the assembly could also cause a kink to propagate. In this case, the MT would be acting as both a signal transducer and a force transducer—just as should be expected, given its nature, and as has recently been confirmed by Donald Ingber and Andrew Maniotis in their landmark study.[7] Since forces can be transmitted from one part of the cell to another in this way, it could as easily work the other way round too: The arrival of a traveling conformational wave could be just enough to trigger a mechanical shift in the cell—giving rise to motion of the cell or its parts. Furthermore, mechanical shifts would likewise cause changes in the electrical status of the MT, a phenomenon known as piezoelectricity—the same thing that powers those annoying, squeaky tunes in "musical" greeting cards and Barney placemats (as readers with children know all too well).

Of course, both these effects come together nowhere more perfectly than in the fourth brain. Action potentials in neurons generate an electric field that influences the propagation of signals in MTs. These signals, in turn, directly affect the physical extension of axons and dendrites, which form the network of the brain. Furthermore, the transfer of a nerve signal from one neuron to another requires the release of chemical neurotransmitters from the end of a transmitting axon onto the tip of a receiving dendrite (the meeting of the two being the synapse). The release involves the physical motion of a neurotransmitter-containing vesicle from the interior of the axon tip to its surface. This physical transport function is likely accomplished—in striking harmony of structure and function—by the end of the MT network that runs along the axon from the cell body.[8] In this fashion, the adaptive computational potential of the microtubular network and its associated proteins would implement the mechanism of Hebbian learning, crucial to the self-organization of intelligence in neural networks.[9]

However, MTs are found in every cell of the body—in every cell of every living creature, in fact. It seems rather odd that they should be structured so as to fit so perfectly the unique requirements of a processing system as extraordinarily complex as the human fourth brain. Surely evolution could not have foreseen these applications when the first axonemes appeared on Earth in the primitive one-celled organisms of four billion years before. It turns out, however, that the tubulin of human MTs are different from the tubulin of any other living creature, and more: The tubulin of the human nervous system is different from the tubulin in MTs elsewhere in the body, and the tubulin of dendrites and axons is different than the tubulin of neuron cell bodies. Neuronal MTs are adapted to their roles.

In general, MTs are unstable structures. They self-assemble and then *disassemble* in a cycle that has long been an object of puzzled fascination for biologists and physicists. (The assembly-disassembly cycle has many features in common with other chemical oscillations, as by now we should expect.) If their function were exclusively skeletal, this might be the means whereby adaptive flexibility could be introduced to the supporting scaffolding of a structure as small as a cell. But in neurons, if we expect MT networks to function as signal propagators, let alone as computational elements, they would need to be much stabler—and they are. Post–self-assembly, *neuronal* MTs undergo a stabilizing process unique to them.

Furthermore, the physical structure of neurons is distinct in having extrusions that may be up to a meter long (motor neuron axons, e.g.). The MTs found in these axons are fantastically longer than those found elsewhere—up to 1 centimeter and composed of circa 100,000 tubulin molecules—roughly 2 billion atoms.[10] Taking into account their diameter (25 nm = 2.5×10^{-9} m), that's the equivalent of a single half-inch-diameter cable over thirteen miles long. To develop such length via self-assembly, they would have to be unusually stable.

TUBULAR!

Let's consider something blindingly obvious about MTs: They're *tubes*.

From the point of view of computation, most of the structures we've examined are either solid arrays (e.g., Hugo de Garis's three-dimensional self-evolving cellular automaton brain) or (mostly) planar arrays (neural networks, cellular automata, ferromagnetic sheets, spin glasses). From a certain point of view, a tube is at once two- and three-dimensional. Unwrapped, it is two dimensional, but one of its two planar dimensions is distinct: The dimension around which the tube is curled in the third dimension. Furthermore, by being curled in three dimensions, it takes two of the boundaries of a plane structure, merges them, and so eliminates both. *Processing or signaling around a tube is continuous.* (This is important, we'll see.) If the plane is *skewed* and then wrapped to form a tube (see figure B–3), a forward or backward bias is introduced (as in a helix) and any process that takes place around the tube might be expected to propagate along it as well. Finally, helical tubes are *chiral*—that is, right- or left-handed, depending on the direction of the twist. Mechanical

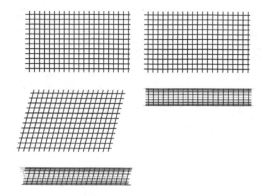

FIGURE B-3   Two-dimensional arrays forming tubes, with and without skew.

strains that have a torsional component will be responded to differently depending on whether the applied torque is aligned with or opposed to the tubular twist.

These are all distinctive features of MTs, and what effect they will have on their processing and/or computational potential is by no means intuitively evident. Tuszynski and his colleagues at the University of Edmonton and abroad have spent the last ten years teasing out the answers to the questions raised by these features. The results are extraordinarily interesting, a testament to the insight of Tuszynski and the genius of nature: Given the exact physical characteristics of tubulin and in its nearest-neighbor interactions, a helical tubular array sustains both signaling and computation in one structure. At physiological temperatures, the MT array appears to assume spin glass–like features, but without sacrificing its communication capacity: a communication system that signals by computing, like an Internet whose wires and cables *are* its computers. The way it seems to work is rather amazing.

The shape of a tubulin depends on whether its mobile (long-distance tunneling) electron is "in" a certain pocket in the α or β subunit, as shown in figure B–4.

The occupied pole of the protein therefore has a slight negative charge as well, offset by a slight positive charge at the other pole. This electron is *very* mobile—tunneling is greatly enhanced by the specific structure of tubulin, by physiological temperatures, and probably by other yet-to-be-flagged charac-

FIGURE B–4   Two conformations of tubulin based on which subunit has the mobile electron.

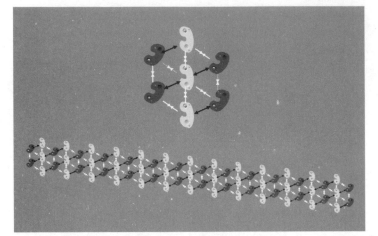

FIGURE B–5   (top) Favored tubulin neighborhood pattern. (bottom) When
linked, the neighborhoods form a consistent pattern, but with a skew. When wrapped,
the ends will be inconsistent with the rest of the array. Net repulsion is shown by
black arrows, net attraction by white.

teristics of its "niche"—and so switches pockets readily. Tubulin responds by
changing its conformation with astounding speed.

As mentioned before, because of the conformational change caused by
the change in electron position, the two opposite dipole states are not mirror
images of one another. The actual spatial direction (positive pole to negative
pole) of an α dipole (a tubulin with an α-subunit electron) is not the precise opposite of a β dipole—it's off the vertical by 29 degrees. The neighborhood for
each dipole is hexagonal, because of the way that tubulins stack. But it's not a
perfectly symmetrical hexagon, it's skewed. This results in a distinct alignment
of the dipoles relative to one another, with some attracting and others repelling, but not in a way that is as self-evident as with square arrays of perfectly
vertical dipoles. When all of these factors are actually calculated, the result is a
distinct favored (lowest-energy) configuration for the six tubulins that form a
neighborhood. (See figure B–5, top.) This configuration happens as well to allow adjacent skewed hexagons to abut one another consistently. (See figure B–
5, bottom.)

However, it also favors a wrapped, triple-helical structure that is thirteen
dimers around. As a result, there's a "seam." But this seam is anything but a
flaw. The seam means that *at least one of the thirteen around-the-tube neighborhood relations must be inconsistent with a perfect low-energy state.* Or, to
put it differently, with respect to any one circuit around the tube, there are thirteen equally low energy states, and these are the lowest possible. The energy
landscape has thirteen valleys. The inconsistency of states around the circumference of the MT (any closed path, no matter how complex) means that *in this
direction there is a degree of frustration as defined for spin glasses.*

The Tuszynski model suggests that at physiological temperatures; and because of the precise parameters involved in the relations among tubulins in a

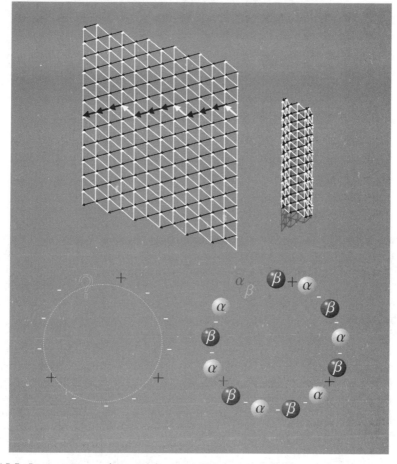

FIGURE B–6   Circumferential frustration in a microtubule. (top left) Arrows represent forces between dipole sites located at the intersections of the lattice. Black arrows are repulsive, white attractive. (top right) The same lattice, wrapped, shows left-handed (black) and right-handed (white) helical lines that are symmetrical. But the purely longitudinal direction has only white lines. The black and white arrows illustrate one sequence with minimal-energy attractive (+) and repulsive (−) dipole forces. When wrapped, these will form a closed circle of neighbor-to-neighbor forces. (bottom left) The same arrangement of minimal-energy attractive (+) and repulsive (−) dipole forces is shown here forming a closed circle—but frustration ensues as no arrangement is perfectly consistent. (bottom right) The arrangement of tubulin states that would correspond to the frustrated lattice structure and the particular closed circle path indicated previously.

neighborhood; and because these parameters favor an odd-numbered (thirteen) unit circumference with skewed hexagons forming a helical tube, the MTs of the human nervous system can function simultaneously as *electromechanical signal transducers (longitudinally) and as spin glass–type parallel computers (circumferentially)*. Figure B–6 summarizes these features schematically.

Appendix C

COHERENCE AND DECOHERENCE

In the double-slit experiment, a single particle is, in some way, "interfering with itself." Since the particle is at the same time a wave, its "parts" have the same source and emerge in "sync," like two waves produced out of one by a double-slit barrier in a water tank. Not only do the "parts" have the same velocity and wavelength, they are "in phase": When one peaks the other does. On the other hand, this doesn't quite make sense. After all, a pattern is recognizable as interference only after many strikes. So, in some sense, the underlying waves belong to all the particles at once. Therefore, even the first can show evidence of the waves' presence. In short, not only are particles also waves, one particle can also be many particles.

Now, the specific pattern of interference caused by two (or more) waves in a superposition depends on what is called their "relative phase." If two waves of the same wavelength are perfectly in phase, peak lines up with peak and trough with trough and the superposition consists of one wave with same wavelength as its two components, but with peaks that are twice as high and troughs that are twice as deep. If two such waves are perfectly out of phase, peak lines up with trough and trough with peak and the superposition consists of nothing—the waves cancel each other out. If the waves are neither perfectly in nor out of phase, the height of their combination can be anything in between, depending on the so-called phase angle between them. (So called because wavelengths are often measured in units of 2π, as angles are.)

Figure C–1 shows three waves of the same wavelength (but different heights, so you can tell them apart), perfectly in phase.

Figure C–2 shows the same waves but at three different phases. They mostly cancel.

The same principles hold for waves in two (or three) dimensions and of various shapes. Figure C–3a shows a radial wave emanating from a point source. Figure C–3b shows what ones sees if *five* identical such waves are superposed in phase. Figure C–3c shows the same five waves but all out of phase.

Naturally, when superposed waves are not identical, their interference pattern is more complicated.

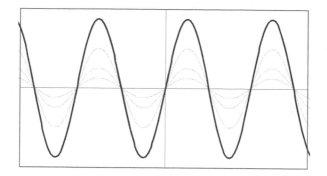

FIGURE C-1   Three waves in phase combine to form a wave of larger amplitude. The three waves are shown as light lines superimposed on one another but not added. The heavy line shows the same three waves added to form a single wave.

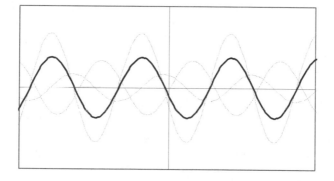

FIGURE C-2   Same three waves as in figure C-1, but not in phase.

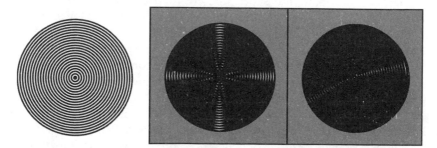

FIGURE C-3   Destruction of interference by phase difference. (left) A radial wave. (middle) Five identical radial waves, superposed in phase. (right) Five identical radial waves, superposed out of phase.

COHERENCE

Most important in the world of quantum effects is the fact that quantum probability amplitudes are not static patterns. They change in time, and these changes are not at all what we might expect. For example, suppose we look at the wave pattern for a free electron (with a known and fixed mass) that we know to have a certain narrow range of possible locations and a certain complementarily narrow range of possible momenta. Suppose that the product of these two ranges—the *uncertainty*—is the minimum physically allowed. Suppose further that the distributions of both momentum and position form a bell-shaped curve—a common enough distribution in natural systems. We place our electron in a perfectly insulated environment and leave it alone for a while.

Intuition suggests that since the electron has been left undisturbed, when we return its wave will have the same shape. But it doesn't. Even so simple a probability distribution as this will *spontaneously* change over time. A bell-shaped curve is not a perfectly *coherent* formation, and like a waistline, it progressively widens as time passes, all by itself.

There are, however, certain very important quantum waves ("eigenfunctions") that do not change their shape over time. Like standing waves on a violin string—the fundamental note or its overtones—each peak and trough on the wave simply grows and shrinks in place. The wave is always the same shape, but its vertical scale varies between its positive maximum and negative minimum. Halfway through every up-and-down cycle, it looks like its mirror image. This kind of change has *no effect whatsoever in the physical world*. The probabilities it yields for finding the particle at any place is the square of the *relative* height or depth, and this never changes even though the underlying probability amplitudes are cycling in time. By "relative" height or depth, we simply mean that whatever the probability amplitudes are, they are proportionally rescaled so the sum of their squares—that is, the sum of all probabilities—equals 1.

Each discrete, quantized energy value of any quantum system has its own such stable pattern *but with its own unique rate of going up and down*. When more than one such pattern is superimposed, therefore, the combined up-and-down motions at *different rates* cause the combined wave to no longer maintain fixed relative values at each point. The probabilities of finding it a given location therefore vary in time. The bell-shaped curve is a typical example of such a combination of eigenfunctions whose overall shape continuously changes.

Now, it is possible to construct such combined waves in such a way that they both maintain their shape forever *and move in space*. Such traveling combined waves are termed "coherent." To create a coherent wave function for a large number of particles is, as you might imagine, extremely difficult.

Here is a striking example of coherence and of how difficult it is to maintain. In the R. L. Pfleegor and L. Mandel experiment of 1967,[1] two separate laser sources of photons with the same wavelength were aimed toward a target but slightly inward, so as to intersect. *There were no barriers*. Pfleegor and Mandel tuned the two separate lasers to the same phase (like tuning two strings on a guitar to the same note) but could keep them tuned to each other only for 20 millionths of a second, since it is so hard to do. When the lasers were thus in phase, a uniform interference pattern appeared on the screen behind their point of intersection. The traveling, superposed waves of light from the two lasers formed a single coherent beam.

They then reduced the intensity of both lasers until no more than one photon at a time, *total*, was emitted—perhaps from the one laser, perhaps from the other. But so long as the tuning of the lasers left any *potential* photons coherent, an interference pattern still appeared during the 20-microsecond window of opportunity. *"[T]he single quantum,"* in other words, *"[wa]s being 'co-produced' in the two different lasers,"* because "[q]uantum theory does not even carry within it the notion of each photon's being produced in a particular laser."[2] Furthermore, there is no limit whatsoever on how far apart the two lasers can be.

The Pfleegor-Mandel experiment not only illustrates the centrality of coherence in quantum mechanics, it also is an example of how the strangest quantum phenomena emerge only when many particles form a coherently superposed whole. Furthermore, the most useful quantum phenomena, including eerily powerful computational capacities, emerge only when coherent superpositions can be produced and maintained for a (relatively) long time.

To create a coherent superposition among a few photons of light, with its relatively long wavelengths, requires extraordinary feats of intelligence, deliberation, and engineering, all building on an ever greater accumulation of knowledge—at least ten thousand years' worth of known human history. To create a coherent superposition among the waves of everyday matter, composed of billions of quanta whose wavelengths are incredibly short, is beyond imagining except in exceptional circumstances. Coherent matter waves have been achieved in the lab only in the last five years. We therefore almost never see quantum interference effects at the scale of everyday life, hence none of the quantum weirdness associated with interfering probability amplitudes. What destroys quantum behavior at everyday scales is decoherence:

> . . . [t]he most worrying difficulty . . . is . . . the problem of *macroscopic interferences* [everyday-scale quantum behavior], which are . . . predicted . . . and practically never observed. . . . [T]his problem led to the idea of decoherence, which is certainly the most important discovery of the modern interpretation.
>
> [D]ecoherence . . . is . . . not a given intrinsic property of nature, but rather a dynamical effect requiring some time to become effective. But how much time? . . . [3]

And why? The answers are:

> . . . [A] very short time. . . . The effect is very efficient, and a few . . . molecules . . . or a few photons . . . can produce significant decoherence . . . or a few phonons [traveling vibrations] that are produced through friction, or by a few electrons. . . . Wave functions of a macroscopic object are extremely sensitive and macroscopic interferences are therefore never observed except under very special conditions. . . .

Decoherence is certainly the most efficient and rapid effect to exist in macroscopic physics. When this simple and basic fact was recognized, one at last had the key to the problem . . . [of absent macroscopic super-positions].[4]

The rapid destruction of quantum interference by even a minute amount of extraneous intrusion from the environment—the tiniest vibration caused by heat, even the *internal* heat of a molecule, or a couple of incidentally tunneled electrons—makes it impossible for huge ensembles of matter at room temperature to sustain a coherent state for meaningful times. For quantum effects to make their presence known in some fashion in systems as complex—and hot—as living matter, environmental agitation must somehow help, not hurt them.

NOTES

Introduction

1. Cited in S. Weinberg, *Dreams of a Final Theory* (New York: Pantheon Books, 1992), p. 255.

2. S. Weinberg, *The First Three Minutes: A Modern View of the Origin of the Universe* (New York: Basic Books, 1977), p. 154.

3. Cited in Weinberg, *Dreams*, p. 256.

4. Abraham Pais, *Niels Bohr's Times in Physics, Philosophy, and Polity* (Oxford: Clarendon Press, 1991), p. 299; and W. Pauli, L. Rosenfeld, and V. Weisskopf, eds., *Niels Bohr and the Development of Physics* (New York: McGraw-Hill, 1955), p. 349.

5. R. Feynman, *The Character of Physical Law* (New York: Modern Library, 1999), p. 9.

Chapter 1

1. M. Gardner, *The Unexpected Hanging and Other Mathematical Diversions: A Classic Collection of Puzzles and Games from Scientific American*, 2nd ed. (Chicago: University of Chicago Press, 1991). In 1961, Donald Michie, a biologist (not a computer scientist) at the University of Edinburgh, invented a device that learned by itself how to play tic-tac-toe. "MENACE (Matchbox Educable Naughts and Crosses Engine)," as he called it, was constructed of three hundred matchboxes. Gardner based Hexapawn on MENACE.

2. Ibid.

Chapter 2

1. W. S. McCulloch and W. Pitts, "A Logical Calculus of Ideas Immanent in Nervous Activity," *Bulletin of Mathematical Biophysics* 5 (1943): 115–133.

2. D. O. Hebb, *The Organization of Behavior: A Neuropsychological Theory* (New York: Wiley, 1949).

3. F. Rosenblatt, "The Perceptron: A Probabilistic Model for Information Storage and Organization in the Brain," *Psychological Review* 65 (1958): 386–408; F. Rosenblatt, *On the Convergence of Reinforcement Procedures in Simple Perceptrons* (Buffalo, NY: Cornell Aeronautical Laboratory, 1960); F. Rosenblatt, "Perceptron Simulation Experiments," *Proceedings of the Institute of Radio Engineers* 48 (1960): 301–309; and F. Rosenblatt, *Principles of Neurodynamics*. (Washington, D.C.: Spartan Books, 1962).

4. P. McCorduck, *Machines Who Think: A Personal Inquiry into the History and Prospects of Artificial Intelligence* (San Francisco: W. H. Freeman and Company, 1979), pp. 87–88.

5. Ibid.

6. Ibid.

7. J. von Neumann, "The General and Logical Theory of Automata," in *Cerebral Mechanisms in Behavior*, ed. L. A. Jeffress (New York: Wiley, 1951).

8. Cited in McCorduck, *Machines Who Think*. Emphasis in original.

9. Ibid., p. 88.

Chapter 3

1. P. McCorduck, *Machines Who Think: A Personal Inquiry into the History and Prospects of Artificial Intelligence* (San Francisco: W. H. Freeman and Company, 1979), p. 88.

2. Ibid., p. 87.

3. Ibid., p. 86.

4. M. Minsky and S. Papert, *Perceptrons: An Introduction to Computational Geometry*, 1st ed. (Cambridge, MA: MIT Press, 1969), p. 4.

5. J. A. Anderson and E. Rosenfeld, eds., *Talking Nets: An Oral History of Neural Networks* (Cambridge, MA: Bradford Books MIT Press, 1998), p. 60.

6. McCorduck, *Machines Who Think*.

7. Minsky and Papert, *Perceptrons*, pp. 231, 232.

8. Cited in Anderson and Rosenfeld, *Talking Nets*, p. 60.

9. P. S. Churchland and T. J. Sejnowski, *The Computational Brain*, 2nd ed. (Cambridge, MA: MIT Press, 1993), p. 109.

10. D. S. Touretzky and D. A. Pomerleau, "What's Hidden in the Hidden Layers?" *Byte* 14 (1989): 227–233.

11. I am not addressing the controversial claims of Noam Chomsky that language per se is a "top-down" "expert system" that we inherit, its rules essentially programmed in place. Although his hypothesis is appearing increasingly dubious in light of the degree of language acquisition displayed by animals, as also by artificial neural networks, even were Chomsky correct, there has never been any question that children learn a *specific* language (e.g., English, Chinese) by imitation and improve it by trial and error. A particularly nice demonstration of how advanced neural networks are in acquiring language in toto without preprogramming—and a particularly nice skewering all at once of the innate-language, postmodern, and deconstructionism fads—is the paper "On the Simulation of Postmodernism and Mental Debility Using Recursive Transition Networks," by Andrew C. Bulhak of the Department of Computer Science at Monash University in Australia (Technical Report 96/254, 1996). Readers may freshly generate their own papers at www.elsewhere.org/cgi/bin/postmodern. The network generates, among other things, full-length fully footnoted and referenced academic/literary papers. An example paragraph:

> "Society is unattainable," says Baudrillard. Derrida uses the term "subsemiotic materialism" to denote the bridge between class and culture. In a sense, the subject is contextualised into a deconstructivist capital-

ism that includes consciousness as a totality. The opening/closing distinction which is a central theme of Virtual Light emerges again in The Burning Chrome.

12. G. Farmelo, "To Err Is Human," review of *Scientific Blunders* by R. Youngson, *New Scientist,* 28 November 1998.

Chapter 4

1. P. Werbos, "Beyond Regression: New Tools for Prediction and Analysis in the Behavioral Sciences," Ph.D. diss. (Cambridge, MA: Harvard University, 1974). It was also proved in works by Y. LeCun, "Une procedure d'apprentissage pour reseau à seuil assymetrique," *Cognitiva* 85 (1985): 599–604, and D. B. Parker, *Learning-Logic: Casting the Cortex of the Human Brain in Silicon* (Cambridge, MA: Center for Computational Research in Economics and Management Science, MIT, 1985).

2. D. E. Rumelhart, G. E. Hinton, and R.J. Williams,"Learning Representations by Back-Propagating Errors," *Nature* 323 (1986): 533–536.

3. J. E. Collard, "A B-P ANN Commodity Trader," in *Advances in Neural Information Processing Systems,* ed. R. P. Lippman, J. E. Moody, and D. S. Touretzky (San Mateo, CA: Morgan Kaufmann Publishers, 1991), pp. 551–556.

4. T. J. Sejnowski and C. R. Rosenberg, "Parallel Networks That Learn to Pronounce English Text," *Complex Systems* 1 (1987): 145–168.

5. J. Stanley, *Introduction to Neural Networks,* 2nd ed., ed. S. Luedeking (Sierra Madre, CA: Scientific Software, 1988), pp. 123, 126.

Chapter 5

1. T. Kohonen, *Self-Organizing Maps,* 2nd ed. (New York: Springer Verlag, 1997), p. 63.

2. T. A. Anderson and E. Rosenfeld, *Talking Nets: An Oral History of Neural Networks* (Cambridge, MA: Bradford Books MIT Press, 1998), pp. 156, 157.

3. G. M. Laskoski, "Addressing Scaling and Plasticity Problems with a Biologically Motivated Self-Organizing Network," in *International Joint Conference on Neural Networks* (San Diego, CA: IEEE, 1990).

4. M. Cottrell, J. C. Fort, and G. Pages, *Technical Report* (Paris: University of Paris, 1994); and Kohonen, *Self-Organizing Maps,* p. ix.

Chapter 6

1. J. J. Hopfield, "Neural Networks and Physical Systems with Emergent Collective Computational Abilities," *Proceedings of the National Academy of Sciences of the USA* 79 (1982): 2554–2558.

2. Jack Cowans cited in J. A. Anderson and E. Rosenfeld, eds., *Talking Nets: An Oral History of Neural Networks* (Cambridge, MA: Bradford Books MIT Press, 1998), p. 113.

3. Aristotle, *On Poetics (de Poetica),* 1st ed. *Great Books of the Western World,* ed. R. M. Hutchins, vol. 9 (Chicago: University of Chicago Press, 1952), p. 1458b.

4. Cited in C. Sykes, ed., *No Ordinary Genius* (New York: W.W. Norton & Company, 1994), pp. 18–19.

5. P. Hoffman, *The Man Who Only Loved Numbers. The Story of Paul Erdös and the Search for Mathematical Truth* (New York: Hyperion, 1998), p. 7.

6. C. G. Jung, *The Psychology of Dementia Praecox,* 1st ed. (New York and Washington: Nervous and Mental Disease Publishing Company, 1936), p. 104.

7. K. R. Jamison, *An Unquiet Mind,* 1st ed. (New York: Knopf, 1996), pp. 4–6.

8. F. C. Hoppensteadt and E. M. Izhikevich, "Weakly Connected Neural Networks," in *Applied Mathematical Sciences,* vol. 126, ed. J. E. Marsden, L. Sirovich, and F. John (New York: Springer, 1997); F. C. Hoppensteadt and E. M. Izhikevich, "Oscillatory Neurocomputers with Dynamic Connectivity," *Physical Review Letters* 82 (14) (1999): 2983–2986; and F. Hoppensteadt and E. Izhikevich, "Canonical Models in Mathematical Neuroscience," *Documenta Mathematica,* 3 (1998): 593–599.

Chapter 7

1. Cited in P. R. Halmos, "The Legend of John von Neumann," *American Mathematical Monthly* 80 (1973): 382–394.

2. E. Wigner, *The Recollections of Eugene P. Wigner as Told to Andrew Szanton,* 1st ed. (New York: Plenum, 1992).

3. N. A. von Neumann, *John von Neumann: As Seen by His Brother* (Meadowbrook, PA: Private Printing, 1987).

4. J. von Neumann, *Mathematical Foundation of Quantum Mechanics* (Princeton, NJ: Princeton University Press, 1996).

5. L. van Hove, "Von Neumann's Contributions to Quantum Theory," *Bulletin of the American Mathematical Society* 64 (1958): 95–99.

6. T. Toffoli and N. Margolus, *Cellular Automata Machines: A New Environment for Modeling,* 4th ed., ed. D. Gannon, Series in Scientific Computation (Cambridge, MA: MIT Press, 1989), p. 9.

7. Ibid.

8. Ibid.

9. See www-swiss.ai.mit.edu/~switz/amorphous/index.html.

10. M. O. Magnasco, "Chemical Kinetics Is Turing Universal," *Physical Review Letters,* 78, no. 6 (1997): 1190–1193. See also: A. Gilman and J. Ross, "Genetic-Algorithm Selection of a Regulator Structure that Directs Flux in a Simple Metabolic Model," *Biophysical Journal* 69 (1995): 1321; N. F. Hansen and J. Ross, "Lyapunov Functions and Relative Stability in Reaction-Diffusion Systems with Multiple Stationary States," *Journal of Physical Chemistry* 100 (1996): 8040; Y. Hung, I. Schreiber, and J. Ross, "Categorization of Some Oscillatory Enzymatic Reactions," *Journal of Physical Chemistry* 100 (1996): 8556; J. P. Laplante, et al., "Experiments on Pattern Recognition by Chemical Kinetics," *Journal of Physical Chemistry* 99 (1995): 10063; J. Ross, A. P. Arkin, and S. C. Muller, "Experimental Evidence for Turing Structures," *Journal of Physical Chemistry* 99 (1995): 10417; and J. Ross, S. C. Mueller, and C. Vidal, "Chemical Waves," *Science* 240 (1988): 460–465.

11. R. Segev and E. Ben-Jacob, "From Neurons to Brain: Adaptive Self-Wiring of Neurons," *Journal of Complex Systems* 1 (1998): 67–78.

12. M. Baringa, "New Leads to Brain Neuron Regeneration," *Science* 282 (1998): 1018–1019; S. L. Florence, H. B. Taub, and J. H. Kaas, "Large-Scale Sprouting

of Cortical Connections After Peripheral Injury in Adult Macaque Monkeys," *Science* 282 (1998): 1117–1121; F. Gage, "Brain Repair," *New Scientist* 160 (1998): 29; D. H. Lowenstein and J. M. Parent, "Brain, Heal Thyself," *Science* 283 (1999): 1126–1127; and P. Rakic, "Young Neurons for Old Brains?" *Nature Neuroscience*, 1, no. 8 (1998): 645–647.

Chapter 8

1. C. G. Langton, "Self-Reproduction in Cellular Automata," *Physica D* 10 (1984): 135–144.

2. J. R. Koza, *Genetic Programming: A Paradigm for Genetically Breeding Populations of Computer Programs to Solve Problems* (Palo Alto, CA: Stanford University, 1990); and J. R. Koza, *Genetic Programming* (Cambridge, MA: MIT Press, 1992).

3. Cited in G. Taubes, "Computer Design Meets Darwin," *Science* 277 (1997): 1931–1932.

4. Cited in ibid.

5. Cited in C. Davidson, "Creatures from Primordial Silicon," *New Scientist* 156 (1997): 30–34.

6. Ibid.

7. Ibid.

8. foobar.starlab.net/~degaris/

Chapter 9

1. J. Monod, *Beyond Chance and Necessity* (London: The Cornerstone Press, 1974).

2. W. Wenzel and K. Hamacher, "Stochastic Tunneling Approach for Global Minimization of Complex Potential Energy Landscapes," *Physical Review Letters* 82, no. 15 (1999): 3003–3007.

3. J. Brooke, et al., "Quantum Annealing of a Disordered Magnet," *Science* 284 1999: 779–781.

Chapter 10

1. A. Whitaker, *Einstein, Bohr and the Quantum Dilemma* (New York: Cambridge University Press, 1996).

2. A. Einstein, *To Erwin Schroedinger*, 1928, cited in G. Greenstein and A. G. Zajonc, *The Quantum Challenge: Modern Research on the Foundations of Quantum Mechanics* (Sudbury, MA: Jones and Bartlett Publishers, 1997), pp. 88–89.

3. A. Einstein, *To the Nobel Prize Committee*, 1928, cited in A. Foelsing, *Albert Einstein. A Biography* (New York: Penguin Group, 1997), p. 590.

4. Cited in Foelsing, *Albert Einstein.*

5. A. Einstein, *To the Nobel Prize Committee*, 1931, cited in Foelsing, *Albert Einstein*, p. 591.

6. A. Einstein, *Autobiographical Notes*, cited in Foelsing, *Albert Einstein.*

7. A. Einstein, B. Podolsky, and N. Rosen, "Can Quantum-Mechanical Description of Physical Reality Be Considered Complete?" *Physical Review Letters* 47 (1935): 777–780.

8. Ibid.

9. S. Rozenthal, *Niels Bohr: His Life and Work as Seen by his Friends and Colleagues* (Amsterdam: North-Holland, 1967), pp. 128–129, cited in Greenstein and Zajonc, *The Quantum Challenge*, pp. 108–109.

10. Greenstein and Zajonc, *The Quantum Challenge*, p. 108.

11. D. Zohar, *The Quantum Self: Human Nature and Consciousness Defined by the New Physics* (New York: William Morrow, 1990).

12. P. W. Anderson, "More Is Different: Broken Symmetry and the Nature of the Hierarchical Structure of Science," *Science* 177 (1972): 393–396.

13. D. N. Mermin, "Is the Moon There When Nobody Looks? Reality and the Quantum Theory," *Physics Today* (1985): 38–47.

Chapter 11

1. R. Feynman, P. R. B. Leighton, and M. Sands, *The Feynman Lectures on Physics, Volume III*, ed. R.B. Leighton (New York: Addison-Wesley, 1965), p. 1–1.

2. R. Feynman, *The Character of Physical Law* (New York: Modern Library, 1965), p. 123.

3. M. Born, "On the Quantum Mechanics of Impact Processes," *Zeitschrift für Physik* 37 (1926): 863.

4. A. Einstein, "Strahlungsemission und Absorption nach der Quantentheorie" ["Emission and Absorption of Radiation According to the Quantum Theory"], *Verhandlungen der Deutschen Physikalischen Gesellschaft [Proceedings of the German Physical Society]* 18 (1916): 318–323.

5. Popular saying attributed to Ernest Rutherford.

6. Feynman, Leighton, and Sands, *Feynman Lectures*.

7. Cited in A. Foelsing, Albert Einstein, A Biography (New York: Penguin Group, 1997), p. 574; see also A. Einstein, *Ueber den gegenwaertigen Stand der Feldtheorie [On the Present State of Field Theory]*, in *Festschrift für A. Stodola*. 1929: Zurich.

8. Cited in Foelsing, *Albert Einstein*, p. 583.

9. Born, "On the Quantum Mechanics of Impact Processes," p. 864; and M. Born, "The Statistical Interpretation of Quantum Mechanics," *Nobel Lectures, Physics* 3 (1954): 318.

Chapter 12

1. C. Monroe et al., "A 'Schroedinger Cat' Superposition State of an Atom," *Science* 272 (1996): 1131–1135.

2. R. Omnés, *The Interpretation of Quantum Mechanics*, 1st ed. Princeton Series in Physics, eds. Philip W. Anderson, Arthur S. Wightman, and Sam B. Treiman, Vol. 22 (Princeton, NJ: Princeton University Press, 1994), p. 159.

3. M. Born, "On the Quantum Mechanics of Impact Processes," *Zetischrift für Physik* 37 (1926): 864, cited in A. Foelsing, *Albert Einstein: A Biography* (New York: Penguin Group, 1997), p. 583.

Chapter 13

1. A. C. Elitzur and L. Vaidman, "Quantum Mechanical Interaction-free Measurements," *Foundations of Physics* 23 (1993): 987–997.

2. T. Tsegaye et al., "Experimental Demonstration of an Efficient Interaction-free Measurement in a Fabry-Perot Interferometer," *Physical Review A* 57 (1998): 3987–3990.

3. A. G. White et al., "'Interaction-free' Imaging," *Physical Review A* 58 (1998): 605–613.

4. Cited in R. Matthews, "Quantum Detectives," *New Scientist* 58 (1998): 11.

5. White et al., "Interaction-free Imaging."

6. M. Buchanan, "Qubits Go Out for a Spin," *New Scientist* 160 (1998): 35; I. L. Chuang, N. Gershenfeld, and M. Kubinec, "Experimental Implementation of Fast Quantum Searching," *Physical Review Letters* 80, no. 15 (1998): 3408–3411; J. A. Jones, "Fast Searches with Nuclear Magnetic Resonance Computers," *Science* 280 (1998): 229; and G. Taubes, "Putting a Quantum Computer to Work in a Cup of Coffee," *Science* 275 (1997): 307–309.

7. R. Jozsa, *Quantum Effects in Algorithms* (Los Alamos, NM: Los Alamos National Laboratory, 1998). Emphasis added. See also R. Jozsa et al., "Universal Quantum Information Compression," *Physical Review Letters* 81, no. 8 (1998): 1714–1717.

8. Jozsa, *Quantum Effects.*

9. M. Brooks, "Crazy Logic," *New Scientist* 2132 (1998): 38–41; R. Penrose, *The Emperor's New Mind: Concerning Computers, Minds and the Laws of Physics* (New York: Oxford University Press, 1989); R. Penrose, "Mechanisms, Microtubules and the Mind," *Journal of Consciousness Studies* 1, no. 2 (1994): 241–249; R. Penrose, *Shadows of the Mind: A Search for the Missing Science of Consciousness* (New York: Oxford University Press, 1994); R. Penrose, *The Large, the Small and the Human Mind*, ed. M. Longair (New York: Cambridge University Press, 1997); and R. Penrose and S. Hameroff, "What 'Gaps'?—Reply to Grush and Churchland," *Journal of Consciousness Studies* 2, no. 2 (1995): 98–111.

10. But there's a big leap hidden here. The kinds of self-evolving devices that scientists like de Garis are creating will not have evolved quantum computational capacities themselves. These capacities will have been built into the evolving substrate—perhaps a quantum cellular automaton (see, for example, C. Duerr and M. Santha, "A Decision Procedure for Unitary Linear Quantum Cellular Automata," quant-ph/9604007 reprints, 1997)—using an extraordinarily intense application of technology to preserve large-scale coherence. There is no evidence as yet that nature at the intracellular level has figured out a way to do likewise.

11. www.cnn.com/TECH/science/9802/18/swiss.robot/index.html.

12. P. Kelley, "Swiss Scientists Warn of Robot Armageddon," February 18, 1998, CNN Online, www.cnn.com/TECH/science/9802/18/swiss.robot/index.html.

Chapter 14

1. J. Preskill, "Quantum Information and Quantum Computing," Course notes for Physics 229, California Institute of Technology, 1997. See theory.caltech.edu/people/preskill/ph229/#lecture.

2. N. Bohr, *Atomic Physics and Human Knowledge* (New York: Science Editions, 1961).

3. Cited in A. March and I. M. Freeman, *The New World of Physics* (New York: Vintage Books, 1963), p. 143.

4. R. Penrose, *Shadows of the Mind. A Search for the Missing Science of Consciousness* (New York: Oxford University Press, 1994).

5. R. Penrose, *The Emperor's New Mind: Concerning Computers, Minds and the Laws of Physics* (New York: Oxford University Press, 1989), p. 402.

6. R. Penrose and S. Hameroff, "What 'Gaps'?—Reply to Grush and Churchland," *Journal of Consciousness Studies* 2, no. 2 (1995): 98–111.

7. Cited in J. Brockman, *The Third Culture* (New York: Simon & Shuster/Touchstone, 1995), pp. 255–256.

8. Cited in ibid., p. 257.

9. M. E. Kellman, personal communication, 1999. See also M. E. Kellman, "Internal Molecular Motions," in *Encyclopedia of Chemical Physics and Physical Chemistry*, ed. J. H. Moore and N. D. Spencer (London: IOP Publishing, in press).

10. J. Hopfield in *Evolutionary Trends in Physical Sciences*, ed. M. Suzuki and R. Kubo (Berlin: Springer, 1991).

11. Cited in Brockman, *Third Culture*, pp. 250–251, emphasis added.

12. R. D. Terry, "The Pathogenesis of Alzheimer Disease: An Alternative to the Amyloid Hypothesis," *Journal of Neuropathology and Experimental Neurology*, 55 (1996): 1023–1025.

13. J. A. Tuszynski, J. A. Brown, and P. Hawrylak, "Dielectric Polarization, Electrical Conduction, Information Processing and Quantum Computation in Microtubules. Are They Plausible?" *Philosophical Transactions of the Royal Society London* 356 (1998): 1897–1926.

14. Ibid.

15. J. A. Brown and J.A. Tuszynski, "Dipole Interactions in Axonal Microtubules as a Mechanism of Signal Propagation," *Physical Review* E 56 (1997): 5834–5840; see also J. A. Brown and J. A. Tuzynski, "Reflections of Biological Signaling: Electronic Conduction May Be an Important Intracellular Pathway," *Advances in Structural Biology* 5 (1998): 115–125.

16. A. C. Maniotis et al., "Demonstration of Mechanical Connections Between Integrins, Cytoskeletal Filaments and Nucleoplasm That Stabilize Nuclear Structure," *Proceedings of the National Academy of Sciences of the USA* 94 (1997): 849–854.

17. J. Glanz, "Force-Carrying Web Pervades Living Cell," *Science* 276 (1997): 678.

18. S. Chang et al., "Transport and Turnover of Microtubules in Frog Neurons Depend on the Pattern of Axonal Growth," *Journal of Neuroscience* 18 (1998): 821–829.

19. D. Van Vactor, *Nerve Cells on the Go—Harvard Researchers Tie Axon Pathfinding to Cytoskeleton Research*. Harvard Medical School press release, 1999.

20. L. M. Adelman, "Computing with DNA," *Scientific American* (August 1998): 54–61; W. Banzhaf, P. Dittrich, and H. Rauhe, "Emergent Computation by Catalytic Reactions," in *Proceedings of the Fourth Foresight Conference on Molecu-*

lar Nanotechnology, 1995; N. Barkai and S. Leibler, "Robustness in Simple Bio-chemical Networks," *Nature* 387 (1997): 913–917; D. Bray, "Protein Molecules as Computational Elements in Living Cells," *Nature* 376 (1995): 307–312; M. Conrad, *Molecular Computing* (New York: Academic Press, 1990), pp. 235–324; M. Conrad, "Molecular Computer Paradigms," *Computer* 25, no. 11 (1992): 6–9; M. Conrad, "Molecular and Evolutionary Computation: The Tug of War Between Context Freedom and Context Sensitivity," *P4* (1998): 117–129; and K. E. Drexler, *Nanosystems: Molecular Machinery, Manufacturing and Computation* (New York: Wiley, 1992).

21. W. Vater et al., "Behaviour of Individual Microtubules and Microtubule Bundles in Electric Fields," in *Proceedings of the Sixth Foresight Conference on Molecu-lar Nanotechnology,* 1998.

22. S. R. Hameroff and R. C. Watt, "Information Processing in Microtubules," *Jour-nal of Theoretical Biology* 98 (1982): 548–561; S. R. Hameroff, S.A. Smith, and R.C. Watt, "Automaton Model of Dynamic Organization in Microtubules," *An-nals of the New York Academy of Sciences* 466 (1986): 949–952.

23. C. G. Langton, *Artificial Life: Santa Fe Institute Studies in the Science of Com-plexity* (New York: Addison-Wesley, 1988).

24. C. G. Langton, "Studying Artificial Life with Cellular Automata," *Physica D* 10, no. 22 (1986): 120–149.

25. P. J. Cachon and M. Cachon, "The Axopodial Systems of *Radiolaria nasse-laria,*" *Archiv für Protistenkunde* 113 (1971): 80–97; A. Samsonovitch, A. Scott, and S. R. Hameroff, "Acousto-conformational Transitions in Cytoskeletal Microtubules: Implications for Intracellular Information Processing," *Nanobiol-ogy* 1 (1992): 457–468.

26. Brown and Tuszynski, "Dipole Interactions in Axonal Microtubules," p. 5835.

27. Ibid.

28. P. S. Churchland and T. J. Sejnowski, *The Computational Brain,* 2d ed. (Cam-bridge, MA: MIT Press, 1993), pp. 86–87.

29. M. V. Sataric, J. A. Tuszynski, and R. B. Zakula, "Kinklike Excitations as an En-ergy-Transfer Mechanism in Microtubules," *Physical Review* E 48 (1993): 589–597.

30. J. Pfaffman and M. Conrad, "Microtubule Networks as a Medium for Adaptive Information Processing," in *Evolutionary Programming VII,* ed. V. W. Porto, N. Saravanan, D. Waagen, and A. E. Eiben (Berlin: Springer, 1988), pp. 463–472; and J. O. Pfaffman and M. Conrad, "Adaptive Signal Processing in Microtubule Networks: A Computer Model," conference presentation, Quantum Approaches to Consciousness, July 1999, University of Arizona, www.consciousness.ari-zona.edu. Emphasis in original.

Chapter 15

1. P. G. Wolynes, "Spin Glass Ideas and the Protein Folding Problem," in *Spin Glasses and Biology,* ed. D. L. Stein (Singapore: World Scientific Publishing, 1992), pp. 225–226.

2. Ibid; R. H. Austin, "The Spin-Glass Analogy in Protein Dynamics," in *Spin Glasses and Biology,* ed. D. L. Stein (Singapore: World Scientific Publishing, 1992), pp. 179–223; S. Gider et al., "Classical and Quantum Magnetic Phenom-ena in Natural and Artificial Proteins," *Science* 268 (1995): 77–80; H. Kawai, T.

Kikuch, and Y. Okamoto, "A Prediction of Tertiary Structures of Peptide by the Monte Carlo Simulated Annealing Method," *Protein Engineering* 3 (1989): 85–94; C. C. Moser et al., "Nature of Biological Electron Transfer," *Nature* 355 (1992): 796–802; J. K. Shin and M. S. Jhon, "High Directional Monte Carlo Procedure Coupled with the Temperature Heating and Annealing Method to Obtain the Global Energy Minimum Structure of Polypeptides and Proteins," *Biopolymers* 31 (1991): 177–185; A. Sornberger, *Quantum Annealing,* NASA/Fermilab Astrophysics Group, Electronic Preprint, 1998; and P. G. Wolynes, "Dynamic Theory of Processes of Chemical Interest in Condensed Matter," National Center for Supercomputing Applications, Digital Information Services, 1999. www.life.uiuc.edu/biophysics/faculty/wolynes.html

3. Kawai et al., "Prediction of Tertiary Structures"; and Shin and Jhon, "High Directional Monte Carlo Procedure."

4. R. S. Farid, C. C. Moser, and P. L. Dutton, "Electron Transfer in Proteins," *Current Opinion in Structural Biology* 3 (1993): 225.

5. G. Goodno, V. Astinov, and R. J. D. Miller, "Diffractive Optics Based Heterodyne Detected Grating Spectroscopy: Application to the Study of Ultrafast Protein Dynamics," *Journal of Physical Chemistry* B 103 (1999): 603–607.

6. A. A. Stuchebrukhov, "Tunneling Currents in Electron Transfer Reactions in Proteins. II. Calculation of Electronic Superexchange Matrix Element and Tunneling Currents Using Non-orthogonal Basis Sets," *Journal of Chemical Physics* 105 (1996): 10819–10829.

7. I. A. Balabin and J. N. Onuchic, "A New Framework for Electron Transfer Calculation—Beyond the Pathways-like Model," *Journal of Physical Chemistry* B 102 (1998): 7497–7596.

8. Ibid.

9. Ibid.

10. M. Y. Ogawa et al., "Distance Dependence of Intramolecular Electron Transfer Rates Across Oligoprolines," *Journal of Physical Chemistry* 97 (1993): 11456–11463; see also B. Dahiyat, T. J. Meade, and S. L. Mayo, "Long-Range Electron Transfer in Isolated Helices," *Inorganica Chimica Acta* 242–243 (1996): 1–6; and A. Vassilian et al., "Electron Transfer Across Polypetides," *Journal of the American Chemical Society* 112 (1990): 7278–7286.

11. I. A. Balabin and H.N. Onuchic, "Connection between Simple Models and Quantum Chemical Models for Electron-Transfer Tunneling Matrix Element Calculation: A Dyson's Equations-Based Approach," *Journal of Physical Chemistry* 100 (1996): 11573–11580; and D. N. Beratan, J. N. Betts, and J. N. Onuchic, "Protein Electron Transfer Rates Set by the Bridging Secondary and Tertiary Structure," *Science* 252 (1991): 1285–1288.

12. J. D. Cruzan et al., "The Far-infrared Vibration-Rotation-Tunneling Spectrum of the Water Tetramer-d8," *Journal of Chemical Physics* 105 (1996): 6634; J. D. Cruzan et al., "Terahertz Laser VRT Spectrum of the Water Pentamer-d10: Constraints on the Bifurcation Tunneling Dynamics," *Chemical Physics Letters* 292 (1998): 667–676; E. H. T. Olthof et al., "Tunneling Dynamics, Symmetry, and Far-Infrared Spectrum of the Rotating Water Trimer, II. Calculations and Experiments," *Journal of Chemical Physics* 105 (1996): 8051; N. Pugliano and R. J. Saykally, "Measurement of Quantum Tunneling Between Chiral Isomers of the Cyclic Water Trimer," *Science* 257 (1992): 1937–1940; and R. J. Saykally,

"Water Clusters," in *The 1999 McGraw-Hill Yearbook of Science & Technology*, ed. S. P. Parker (New York: McGraw-Hill, 1999).

13. A. Kohen and P. J. Klinman, "Enzyme Catalysis: Beyond Classical Paradigms," *Accounts of Chemical Research* 31 (1998): 397–404.

14. J. Basran, M. J. Sutcliffe, and N. S. Scrutton, "Enzymatic H-transfer Requires Vibration-driven Extreme Tunneling," *Biochemistry* 38 (1999): 3218–3222; I. Daizadeh, E. S. Medvedev, and A. A. Stuchebrukhov, "Effect of Protein Dynamics on Biological Electron Transfer," *Proceedings of the National Academy of Science USA* 94, no. 8 (1997): 3703–3708; and Stuchebrukhov, "Tunneling Currents in Electron Transfer Reactions."

15. I. Daizadeh and A. Stuchebrukhov, "Effects of Protein Dynamics on Biological Electron Transfer," American Physical Society, APS/AAPT Joint Meeting, 1998.

16. Basran et al., "Enzymatic H-transfer."

17. The "speed" at which an electron tunnels remains puzzling—if it may be spoken of at all. The process is either instantaneous for "part of its trajectory" referred to as an "instanton" or in any event faster than light. In 1993 Steinberg, Kwait, and Chiao showed that the total transit time for the tunneling "half" of a split photon was less than for the nontunneling half, that is faster than the speed of light.

18. Wolynes, "Spin Glass Ideas."

19. P. Wolynes and A. Kuki, "Electron Transfer Paths in Protein," National Center for Supercomputing Applications, Digital Information Services, 1998.

20. Austin, "Spin-Glass Analogy."

21. B. J. Bahnson and J. P. Klinman, "Hydrogen Tunneling in Enzyme Catalysis," *Methods in Enzymology* 249 (1995): 373–397; B. J. Bahnson et al., "A Link Between Protein Structure and Enzyme Catalyzed Hydrogen Tunneling," *Proceedings of the National Academy of Science USA* 94, no. 24 (1997): 12797–12802; B. J. Bahnson et al., "Unmasking of Hydrogen Tunneling in the Horse Liver Alcohol Dehydrogenase by Site-directed Mutagenesis," *Biochemistry* 32, no. 27 (1993): 5503–5507; F. Bartl et al., "The FO Complex of the ATP Synthase of Escherichia coli Contains a Proton Pathway with Large Proton Polarizability Caused by Collective Proton Fluctuation," *Biophysics* 68, no. 1 (1995): 104–110; F. Bartl et al., "Proton Relay System in the Active Site of Maltodextrinphosphorylase via Hydrogen Bonds with Large Proton Polarizability: An FT-IR Difference Spectroscopy Study," *European Biophysics* 28, no. 3 (1999): 200–207; K. L. Grant and J. P. Klinman, "Evidence that Both Protium and Deuterium Undergo Significant Tunneling in the Reaction Catalyzed by Bovine Serum Amine Oxidase," *Biochemistry* 28, no. 16 (1989): 6597–6605; W. E. Karsten, C. C. Hwang, and P. F. Cook, "Alpha-secondary Tritium Kinetic Isotope Effects Indicate Hydrogen Tunneling and Coupling Motion Occur in the Oxidation of L-malate by NAD-mali Enzyme," *Biochemistry* 38, no. 14 (1999): 4398–4402; J. P. Klinman, "Quantum Mechanical Effects in Enzyme-Catalysed Hydrogen Transfer Reactions, " *Trends in Biochemical Science* 12, no. 9 (1989): 368–373; A. Kohen et al., "Enzyme Dynamics and Hydrogen Tunneling in a Thermophilic Alcohol Dehydrogenase," *Nature* 399 (1999): 496–499; and A. Kohen, T. Jonsson, and J. P. Klinman, "Effects of Protein Glycosylation on Catalysis: Changes in Hydrogen Tunneling and Enthalpy of Activation on the Glucose Oxidase Reaction," *Biochemistry* 36, no. 22 (1997): 2603–2611.

22. Basran et al., "Enzymatic H-transfer."

23. Daizadeh and Stuchebrukhov, "Effects of Protein Dynamics."

24. H. Linke, University of New South Wales, 1999,
www.phys.unsw.edu.au/STAFF/RESEARCH/linke.html; and H. Linke, W. Sheng,
A. Löfgren, Hongqi Xu, P. Omling, and P. E. Lindelof, "A Quantum Dot
Ratchet: Experiment and Theory," *Europhysics Letters* 44, no. 3 (1998): 341–
347.

25. Kohn et al., "Enzyme Dynamics and Hydrogen Tunneling."

26. Ibid.

27. K. Drukker, S. D. Leeuw, and S. Hammes-Schiffer, "Proton Transport along
Water Chains in an Electric Field," *Journal of Chemical Physics* 108 (1998):
6799; S. Hammes-Schiffer, "Multiconfigurational Molecular Dynamics with
Quantum Transitions: Multiple Proton Transfer Reactions," *Journal of Chemical Physics* 105 (1996): 2236–2246; and D. J. Wales and T. R. Walsh, "Theoretical Study of the Water Pentamer," *Journal of Chemical Physics* 105 (1996):
6597.

28. D. Beratan and S. Skourtis, "Electron Transfer Mechanisms," *Current Opinion
in Chemical Biology* 2, no. 2 (1998): 235–243.

29. J. Barton, *DNA-mediated Electron Transfer Chemistry* (Pasadena, CA: Institute
of Technology, 1999). Emphasis added.

30. Cited in A. Coghlan, "Electric DNA," *New Scientist* 161 (1999): 19.

31. J. Monad, *Beyond Chance and Necessity* (London: The Cornerstone Press,
1974).

32. W. G. Cooper, "Roles of Evolution, Quantum Mechanics and Point Mutations
in Origins of Cancer," *Cancer Biochemistry and Biophysics* 13, no. 3 (1993):
147–70.

33. W. G. Cooper, "T4 Phage Evolution Data in Terms of a Time-dependent Topal-
Fresco Mechanism," *Biochemical Genetics* 32, no. 11–12 (1994): 383–395.

34. M. Conrad, "Superinformation Processing: The Feasibility of Proton Superflow
in the Living State," in *Molecular and Biological Physics of Living Systems*, ed.
R. K. Mishra (Dordrecht, Netherlands: Kluwer Academic Publishing, 1990), pp.
159–174.

Chapter 16

1. J. Ford, "How Random Is a Coin Toss?" *Physics Today* (1983): 46.

2. F. Capra, *The Tao of Physics*, 3rd, updated ed. (Boston: Shambhala, 1991); I.
Marshall and D. Zohar, *Who's Afraid of Schroedinger's Cat? All the New Science Ideas You Need to Keep Up with the New Thinking* (New York: William
Morrow & Company, 1997); F. A. Wolf, *Star Wave. Mind, Consciousness, and
Quantum Physics* (New York: Macmillian, 1984); and D. Zohar, *The Quantum
Self: Human Nature and Consciousness Defined by the New Physics* (New
York: William Morrow, 1990).

3. N. Birbaumer, H. Flor, W. Lutzenberger, and T. Elbert, "Chaos and Order in the
Human Brain," *Electroencephalography and Clinical Neurophysiology Supplement* 44 (1995): 450–459; S. P. Layne, G. Mayer-Kress, and J. Holzfuss, eds.,
"Problems Associated with Dimensional Analysis of Electro-Encephalogram
Data," *Dimensions and Entropies in Chaotic Systems,* ed. G. Mayer-Kress, Vol.
32 (Berlin: Springer-Verlag, 1986), pp. 246–256; G. Mayer-Kress, "Nonlinear

Dynamics, Chaos, and Brain Signals," *Psychophysiology* 25 (1988): 419; G. Mayer-Kress, "Non-Linear Mechanisms in the Brain," *Zeitschrift für Naturforschung* 53c, no. 7/8 (1998): 667–685; G. Mayer-Kress, C. Barczys, and W. J. Freeman, "Attractor Reconstruction from Event-Related Multi-Electrode EEG-Data," in *Mathematical Approaches to Brain Functioning Diagnostics*, ed. A. V. Holde (Hong Kong: World Scientific, 1991); G. Mayer-Kress and J. Holzfuss, "Analysis of the Human Electroencephalogram with Methods from Nonlinear Dynamics," in *Proceedings of Temporal Disorder in Human Oscillatory Systems*, ed. M.C. Mackey (Berlin: Springer-Verlag, 1987), pp. 57–68; G. Mayer-Kress and S. P. Layne, "Dimensionality of the Human Electroencephalogram," *Perspectives in Biological Dynamics and Theoretical Medicine* 504 (1987): 62–86; G. Mayer-Kress et al., "Dimensional Analysis of Nonlinear Oscillations in Brain, Heart and Muscle," *Mathematical Biosciences* 90 (1988): 155–182; and H. T. Schupp et al., "Neurophysiological Differences between Perception and Imagery," *Cognitive Brain Research* 2 (1994): 77–86.

4. H. Liljenstrom, "Oscillations and Associative Memory: Brain and Model," in *Brain and Mind–Biology Transactions of the Danish Academy of Sciences and Letters*, ed. R. Cotterill (Copenhagen: Danish Academy of Science and Letters, 1994), pp. 109–123; H. Liljenstrom, "Modeling the Dynamics of Olfactory Cortex Using Simplified Networks Units and Realistic Architecture," *International Systems Journal* 2 (1991): 1–15; H. Liljenstrom and X. Wu, "Noise-enhanced Performance in a Cortical Associative Memory Model," *Neural Systems* 6 (1995): 19–29; and J. J. Wright and D. T. J. Liley, "Dynamics of the Brain at Global and Microscopic Scales: Neural Networks and the EEG," *Behavior and Brain Sciences* 19 (1996): 285–320.

5. H. Liljenstrom, "Modeling the Dynamics of Olfactory Cortex." See also H. Liljenstrom and M. Hasselmo, "Cholinergic Modulation of Cortical Oscillatory Dynamics," *Neurophysiology* 74 (1995): 288–297.

6. W. Freeman, "Chaos in the CNS: Theory and Practice," in *Neural Modeling and Neural Networks,* ed. F. Ventriglia (New York: Pergamon Press, 1989), pp. 185–216; W. J. Freeman, "The Physiology of Perception," *Scientific American* 264 (1991): 78–85.

7. Liljenstrom, "Modeling the Dynamics of Olfactory Cortex"; T. Wilhelmsson and H. Liljenstrom, "Simulating the Complex Dynamics of a Brain Structure," in *PDC Center for Parallel Computers Progress Report*, ed. F. Hedman, P. Hammarlund, and J. Oppelstrup (Stockholm, Sweden: KTH, 1994), pp. 24–28; T. Wilhelmsson and H. Liljenstrom, *Simulating the dynamics of olfactory cortex on the CM200*, 1994; and X. Wu and H. Liljenstrom, "Regulating the Nonlinear Dynamics of Olfactory Cortex," *Network* 5 (1994): 47–60.

8. M. G. Rosenblum, A. S. Pikovsky, and J. Kurths, "Phase Synchronization of Chaotic Oscillators," *Physical Review Letters* 76, no. 11 (1996): 1804–1807.

9. D. J. Albers and J. C. Sprott, "Routes to Chaos in Neural Networks with Random Weights," *International Journal of Bifurcation and Chaos* 8 (1998): 1463–1478.

10. P. R. Halmos, "The Legend of John von Neumann," *American Mathematical Monthly* 80 (19): 382–394.

11. T. O. Scheper, N. Crook, and C. Dobbyn, "Chaos as a Desirable Stable State of Artificial Neural Networks," in *Proceedings of the International ICSC Symposium on Neural Computation* (Vienna: Technical University of Vienna, 1998).

12. S. Sinha and W. L. Ditto, "Dynamics Based Computation," *Physical Review Letters* 81, no. 10 (1998): 2156–2159.

13. M. Makishima and T. Shimizu, "Wandering Motion and Co-Operative Phenomena in a Chaotic Neural Network," *International Journal of Bifurcation and Chaos* 8, no. 5 (1998): 891–898.

14. Birbaumer et al., "Chaos and Order"; N. Birbaumer, "Perception of Music and Dimensional Complexity of Brain Activity," *International Journal of Bifurcations and Chaos* 6, no. 2 (1996): 267–278; Layne et al., "Problems Associated with Dimensional Analysis"; Mayer-Kress and Layne, "Dimensionality of the Human Encephalogram"; Mayer-Kress et al., "Attractor Reconstruction from Event-Related Multi-Electrode EEG-Data"; Mayer-Kress et al., "Dimensionality Analysis of Nonlinear Oscillations"; and Schupp et al., "Neurophysiological Differences."

15. See, for example, B. B. Mandelbrot, "Price Change and Scaling in Economics," in *The Fractal Geometry of Nature* (New York: W.H. Freeman & Co., 1983), pp. 334–340.

16. G. Mayer-Kress, "Global Brains as Paradigm for a Complex Adaptive World," in *Evolving Complexity: Challenges to Society, Economy, and the Individual* (Dallas: University of Texas, 1994); G. Mayer-Kress, "Messy Futures and Global Brains," in *Predictability of Complex Dynamical Systems*, ed. Y. A. Kravtsov and J. B. Kadtke (Berlin: Springer-Verlag, 1996); G. Mayer-Kress and C. Barczys, "The Global Brain as an Emergent Structure form the Worldwide Computing Network, and Its Implications for Modeling," *Information Society* 11, no. 1 (1995): 1–28.

17. B. Drossel, "Simple Model for the Formation of a Complex Organism," *Physical Review Letters* 82, no. 25 (1999): 5144–5147.

18. S. Tomsovic and E. J. Heller, "Long-time Semiclassical Dynamics of Chaos: The Stadium Billiard," *Physical Review E* 47 (1993): 282–284.

19. R. V. Jensen, "Chaos," in *Encyclopedia of Modern Physics*, ed. R. A. Meyers (New York: Academic Press, 1990), pp. 69–96, quoted from p. 94.

20. M. E. Kellman, personal communication, 2000. See also J. P. Rose and M. E. Kellman, "Spectral Patterns of Chaotic Acetylene," *Journal of Physical Chemistry* A, in press.

21. G. Casati, G. Maspero, and D. L. Shepelyansky, "Quantum Poincaré Recurrences," *Physical Review Letters* 82, no. 3 (1999): 524–527.

22. J. S. Hersch, M. R. Haggerty, and E. J. Heller, "Diffractive Orbits in an Open Microwave Billiards," *Physical Review Letters* 83, no. 25 (1999): 5342–5345.

23. H. Amman et al., "Quantum Delta-Kicked Rotor: Experimental Observation of Decoherence," *Physical Review Letters* 80, no. 19 (1998): 4111–4115.

24. J. P. Keating et al., *Disordered Systems and Quantum Chaos: Report from the Organisers* (New York: Cambridge University, 1997). Emphasis added.

Chapter 17

1. A. S. Eddington, *The Nature of the Physical World* (Cambridge: Cambridge University Press, 1928), p. 211.

2. C. P. Collier et al., "Electronically Configurable Molecular-Based Logic Gates," *Science* 285 (1999): 391–398.

3. W. T. Buttler et al., "Practical Free-Space Quantum Key Distribution over 1 km," *Physical Review Letters* 81, no. 15 (1998): 3283–3286.

4. C. Monroe et al., "A 'Schroedinger Cat' Superposition State of an Atom," *Science* 272, no. 5265 (1996): 1131–1135.

5. J. Polkinghorne, *The Faith of a Physicist* (Princeton, NJ: Princeton University Press, 1994).

6. Cited in M. Matsumara, ed., *Voices for Evolution* (El Cerrito, CA: National Center for Science Education, 1995), p. 33.

7. T. H. Huxley, *Collected Essays*, vol. 2, *Darwinia: The Origin of the Species* (London: Macmillan, 1893), p. 52.

8. S. Begley, "Science Finds God," in *Newsweek* (1998): 46–51. See also G. Easterbrook, "Science and God: A Warming Trend?" *Science* 277 (1997): 890–893; and G. Johnson, "Science and Religion: Bridging the Great Divide," in *New York Times*, July 1, 1998; p. F4.

9. Johnson, "Science and Religion."

10. E. J. Larson and L. Witham, "Scientists Are Still Keeping the Faith," *Nature* 386 (1997): 435–436.

11. E. J. Larson and L. Witham, "Leading Scientists Still Reject God," *International Weekly Journal of Science Nature* 394 (1998): 313.

12. Cited in Johnson, "Science and Religion."

13. Cited in Easterbrook, "Science and God."

14. Ibid.

15. Ibid.

16. Ibid.

17. Ibid.

18. Ibid.

19. Ibid.

20. R. Dawkins, *River Out of Eden: A Darwinian View of Life* (New York: Basic Books, 1995), p. 133.

21. M. E. Kellman, personal communication, 2000.

22. A. A. Harkavy, "Speculations Concerning Will and a Local God," in *Human Will: The Search for its Physical Basis* (New York: Peter Lang, 1995).

23. P. S. Churchland and T. J. Sejnowski, *The Computational Brain*, 2d ed. (Cambridge, MA: MIT Press, 1993), pp. 1–2.

Appendix B

1. M. A. Collins et al., *Physical Review B* 19 (1979): 3630.

2. E. W. Montroll, in *Statistical Mechanics*, ed. S. A. Rice, K. F. Freed, and J. C. Light (Chicago: University of Chicago, 1972).

3. K. C. Chou, C. T. Zhang, and G. M. Maggiora, *Biopolymers* 34 (1994): 143.

4. M. V. Sataric, J. A. Tuszynski, and R. B. Zakula, "Kinklike Excitations as an Energy-Transfer Mechanism in Microtubules," *Physical Review E* 48 (1993): 589–597; M. V. Sataric, R. B. Zakula, and J. A. Tuszynski, "A Model of the Energy

Transfer Mechanism in Microtubules Involving a Single Solition," *Nanobiology* 1 (1992): 445–450.

5. M. V. Sataric, R. B. Zakula, Z. Ivic, and J. A. Tuszynski, "Influence of a Solitonic Mechanism on the Process of Chemical Catalysts," *Journal of Molecular Electronics* 7 (1991): 39–46.

6. P. Woafo, G. H. Guenoue, and A. S. Bokosah, "Discreteness Effects on Kinklike Excitations in Microtubules," *Physical Review E* 55, no. 1 (1997): 1209–1212.

7. A. C. Maniotis et al., "Demonstration of Mechanical Connections between Integrims, Cytoskeletal Filaments and Nucleoplasms that Stabilize Nuclear Structure," *Proceedings of the National Academy of Sciences of the USA* 94 (1997): 849–854; J. Glanz, "Force-Carrying Web Pervades Living Cell," *Science* 276 (1997): 678.

8. L. A. Segel and H. Parnas, *Biologically Inspired Physics* (New York: Plenum, 1991); and J. A. Tuszynski, J. A. Brown, and P. Hawrylak, "Dielectric Polarization, Electrical Conduction, Information Processing and Quantum Computation in Microtubules. Are They Plausible?" *Philosophical Transactions of the Royal Society, London* 356 (1998): 1897–1926.

9. M. Maletic-Savatic, R. Malinow, and K. Svoboda, "Rapid Dendritic Morphogenesis in CA1 Hippocampal Dendrites Induced by Synaptic Activity," *Science* 283 (1999): 1923–1927; and S. J. Smith, "Dissecting Dendrite Dynamics," *Science* 283 (1999): 1860–1861.

10. J. A. Brown and J. A. Tuszynski, "Dipole Interactions in Axonal Microtubules as a Mechanism of Signal Propagation," *Physical Review E* 56, no. 5 (1997): 5834–5840; Satavic et al., "Kinklike Excitations as an Energy-Transfer Mechanism"; and B. Trpisova and J. A. Tuszynski, "Possible Link between Guanosine 5' Triphosphate Hydrolysis and Solitary Waves in Microtubules," *Physical Review E* 55, no. 3 (1997): 3288–3305.

Appendix C

1. R. L. Pfleegor and L. Mandel, "Interference of Independent Photon Beams," *Physical Review* 159 (1967): 1084–1088.

2. G. Greenstein and A. G. Zajonc, *The Quantum Challenge: Modern Research on the Foundations of Quantum Mechanics* (Sudbury, MA: Jones and Bartlett Publishers, 1997), pp. 45–46.

3. R. Omnés, *Understanding Quantum Mechanics* (Princeton, NJ: Princeton University Press, 1999), p. 76.

4. Ibid., 77.

CREDITS

Figures 5–1 and 5–4(a): Reproduced from *The Neurosciences* by copyright permission of the Rockefeller University Press.

Figure 5–4(b): Reproduced from Lorente de No, R., "The Structure of the Cerebral Cortex," in *Physiology of the Nervous System,* J. F. Fulton, Editor. 1949, Oxford University Press: New York, pp. 288–330, by copyright permission.

Figures 7–4, 7–5, and 7–9(b): Generated by and used with the permission of the Math and Computation Department, MIT Artificial Intelligence Lab.

Figure 7–8(c): From *Exploring Complexity* by Prigogine and Nicolis. © 1989 by Gregoire Nicolis and Ilya Prigogine. Used with permission by W. H. Freeman and Company.

Figure 8–5: Adapted from simulations by and used with the permission of Juha Haataja at the Center for Scientific Computing, Espoo, Finland.

Figures 8–9(a) and 8–9(b): © Hugo de Garis. Reproduced with permission.

Figure 11–10: Reprinted with permission from Tonomura, A., Endo, J. et al, "Demonstration of Single-Electron Build-Up of an Interference Pattern," *American Journal of Physics,* 57(2), 1989, pp. 117–120. © 1989, American Association of Physics Teachers.

Figure 11–13(b): © 1999, W. Blaine Stine, Evanston Northwestern Healthcare Research Institute. Reproduced with permission.

Figures 11–13(c) and 12–1: Courtesy IBM Almaden Research Laboratories.

Figure 14–1: J. R. McIntosh, "The Roles of Microtubules in Chromosome Movement," in *Microtubules,* edited by J. S. Hyams and C. W. Lloyd. © 1994 J. R. McIntosh. Reprinted by permission of Wiley-Liss, Inc., a subsidiary of John Wiley & Sons, Inc.

Figure 14–3(a) and 14–4(c): Gary Borisy, © 1997, Laboratory of Molecular Biology, Northwestern University. Reproduced with permission.

Figure 14–3(b) and 14–4(f): Linda A. Amos, © 1999 MRC Laboratory of Molecular Biology, Cambridge University. Reproduced with permission.

Figure 14–3(d): E. F. Smith and W. S. Sale, "Mechanisms of Flagellar Movement," in *Microtubules,* edited by J. S. Hyams and C. W. Lloyd. © 1994 J. R. McIntosh. Reprinted by permission of Wiley-Liss, Inc., a subsidiary of John Wiley & Sons, Inc.

Figure 14–4(a): J. M. Scholey and R. D. Vale, "Kinesin-based Organelle Transport," in *Microtubules,* edited by J. S. Hyams and C. W. Lloyd. © 1994 J. R. McIntosh. Adapted and reprinted by permission of Wiley-Liss, Inc., a subsidiary of John Wiley & Sons, Inc.

INDEX